HVAC
ELECTRICAL FOR IDIOTS

Brien Hollis

Copyright © 2021 Brien Hollis
All rights reserved
First Edition

NEWMAN SPRINGS PUBLISHING
320 Broad Street
Red Bank, NJ 07701

First originally published by Newman Springs Publishing 2021

ISBN 978-1-63692-902-6 (Paperback)
ISBN 978-1-63692-903-3 (Digital)

Printed in the United States of America

Dedicated to my grandchildren:

Paisley Pillow
Averi Pillow
Ethan Hollis

BEFORE WE GET STARTED

The following page has the four components that are consistent throughout the book. All circuits that are built are called "Series" circuits and must have these four components. In order to use this book to troubleshoot series circuits, you must write these four items in order. They are: Power, Switch, Load, and Common. This order will help you as you build your skills. In order to be a successful troubleshooter, you need to use a procedure that you can mimic over and over for 100 percent accuracy.

Let's build this circuit. Take a piece of 30 Awg insulated wire and wrap it around a nail as many times as you can. Leave approximately three to four inches before you wrap it. When you are finished wrapping the wire around the nail, leave enough wire to connect to tape to the negative side of the battery. Tape a piece of the same size wire to the positive side of the battery. Depending on the size of the nail and the amount of times you wrapped it around the nail may require two batteries. When you put the two wires together ("A" and "B"), we call this closing the switch. At this time the nail will become a magnet. This circuit is how all relays, contactors, and electric motors are built. It is called Electromagnetism.

When you take the two wires apart, we have now opened the switch and the magnetism goes away. This magnetism is how relays, contactors, and many other switches open or close a circuit.

Step down transformers operate with the same magnetism. We will discuss these more later in the book.

What is the MOST important idea to realize in this circuit is which part is **Power**, which one is **Switch**, which one is **Load**, and which item is **Common**.

We need all four of these to have a controlled circuit. If you do not have 30 AWG wire, you can use thermostat wire (leave insulation on) and a 24 Volt step down transformer. Or if you have a contactor, you can take the coil out it and have a premade coil. A cheap extension cord can be used for power to the transformer, but be very careful to only use the 24 Volts to the coil. Do not leave 120 Volts to the transformer uncovered. Tape all exposed on the transformer before plugging in.

Good luck! Let's have some fun.

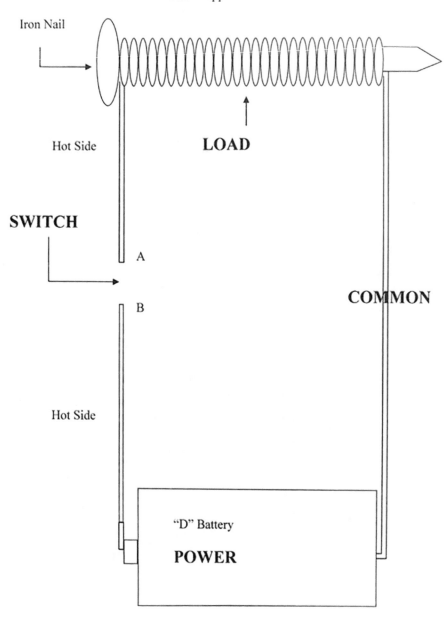

CHAPTER 1

Building Series Circuits

CAUTION: Voltages can cause death. Use caution when testing all electrical circuits.

Often, technicians get overwhelmed with the amount of wiring inside a furnace or air conditioner (A/C) unit. It can be intimidating, but doesn't have to be. What we are actually looking at, is many different circuits combined in parallel. When they are broken into individual circuits, they are quite simple to troubleshoot. It is important to learn that there is a sequence of operation in all heating and cooling equipment. In other words it is important to know; what happens first, what happens second, what happens third, and so on. The sequence of operation will be covered in later chapters for each type of heating and cooling equipment. Please take time to memorize those sequences of operations. When HVAC equipment breaks down, it is usually due to an individual part, a few parts, or the power supply in the circuit.

First, let's understand the basic components of a simple circuit. They are **Power** (supply or power supply), **Switch** (or switches), **Load**, and **Common**. For all intents and purposes we will be discussing single phase alternating current. This is commonly referred to as AC voltage or AC current. Alternating current (AC) does not have a positive or negative but it does have a direction of flow. In an electrical conductor (wire) we are talking about the free electrons that are moving (or flowing) through the wire. Think of this electron flow as water flowing through a hose. As an example, water from a hose will be turning a water wheel (load). The water going to the wheel is under pressure (hot). As the water falls off the side of the water wheel, it is no longer under pressure (common). Electrons will move from the power source, through the wire to the load under pressure (hot), and return through the common.

What is a load? A load can be anything that requires electricity (electrons) to operate. A load could be a motor, a light, heating element, etc. All loads have a resistance value. In other words, the resistance in the load slows the flow of electrons or movement of electricity. The more the resistance, the higher amperage or a larger amount of electrons is required to overcome the resistance created by the load.

Figure 1-1 shows a complete and controlled circuit. It indicates the four basic items needed for that circuit (**Power, Switch, Load**, and **Common**). The lines indicate conductors (wires) that connect each item. Power is leaving the supply to the switch. Then, it continues from the switch and goes to the load. The left and is called the hot side. Any wire before the load is said to be on the hot side. Then, it goes through the load and returns through the common. The right side is the common. Any wire that is located after the load is said to be on the common side.

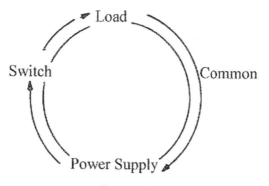

Figure 1-1

To better understand, divide a series circuit into two sides; the hot side and the common side. Drawing a vertical line down through the center will help you understand how one side is used to find missing voltages or problems. Later, we will see how to use each side to determine where a problem exists. This will work with and all electrical circuits or circuit boards that use alternating current. **NOTE; You need your meter to indicate both sides of power before any voltages will show on your meter.** Figure 1-2 shows the four minimum components to have a controlled circuit (**Power**, **Switch**, **Load**, and **Common**). Please take time to study and memorize how these components are used.

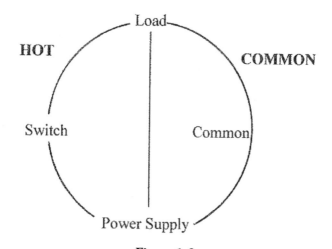

Figure 1-2

In figure 1-2, we have drawn a vertical line showing how the circuit is divided into two sides. The left side is the **hot side** and the right side is the **common side**. Although we might not know the direction of electron flow, the side that contains the switch will be the **hot side**. This means the electricity is flowing from the **power** supply to the **switch**, then to the **load**, then through the load, and then goes back to the power supply through the **common** wire. Are you sensing a theme here (**Power, Switch, Load,** and **Common**)? Regardless if the load is a blower motor, gas valve, or circuit board, these four things never change in the arrangement.

All switches, whether one or multiple, will **always** be located between the power supply and the load. Placing switches on the common side is dangerous. **Never place a switch after the load.** The four items **MUST** stay in exact order. Current (the flow of electrons) can flow in either direction, but do not have polarity (positive or negative).

In circuit boards, the flow of electrons must travel in only one direction. This direction is already given to you by your utility supplier. The black wire entering the equipment is the hot side and the white wire is the common (or neutral). The solid unwrapped wire is the ground and is also a common or neutral. It serves as a pathway of the white common is lost. Electricity flows to ground (mother earth) through the shortest and least resistant path. Once a path is established, all circuits must be wired for that direction of flow. Figure 1-3 shows a correct circuit. Figure 4 shows an incorrect circuit. Please study and understand the differences in these figures. **Remember; NEVER place a switch after a load.**

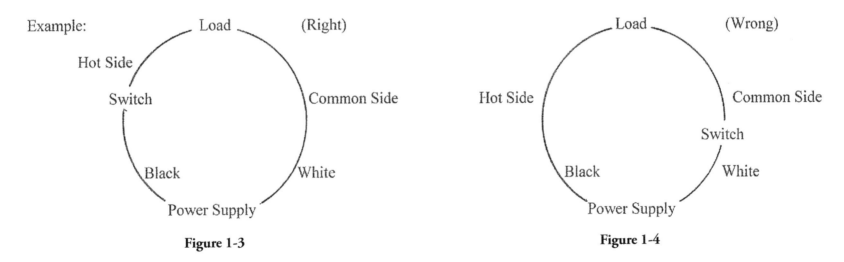

Figure 1-3   **Figure 1-4**

Both examples in figures 1-3 and 1-4 would operate the circuit. However, Figure 1-4 would be hazardous because a switch has been placed after the load. **Remember; NEVER place a switch after a load.** A circuit board will not work with electrons flowing backwards because the diodes on the board will allow electrons to flow in only one direction. Reversing the flow of electrons may cause only a portion of the circuit board to operate correctly or the entire board might not work.

HVAC ELECTRICAL FOR IDIOTS | 9

VOLTAGES AND SEQUENCE OF OPERATIONS

Once the direction of flow has been determined, troubleshooting is a breeze. When wired properly, we do not need to check the direction of flow. If the switch is before the load, then that is the direction of flow or the hot side. All commons are put together. If you look at the back side of a circuit board you will see that all of the commons are connected together. Using any one of those common terminals would be the same for troubleshooting. Just make sure they are the correct voltage. What are the correct voltages used in residential heating and cooling? With forced air furnaces and boilers, voltage readings might vary from 110 volts to 122 volts on the 120 volt portion. This 120 voltage is used for indoor blower motors, draft inducer motors, hot surface igniters, transformers, and etc. The 24 volt portion is considered a control voltage that operates or controls the larger loads and voltages. Measured voltages range from 24 volts to 27 volts. The 24 volt portion operates relays, contactors, and smaller loads.

Most air conditioning compressors and condenser fan motors requires 220 volts and usually measure between 220–223 volts of alternating current. Some units are 208 ac volts, 227ac volts, or 230ac volts. Always look on the data plate located on the unit for the exact voltages required. Use your multimeter and its RMS (root mean square) values. If your multimeter does not have true RMS, then multiply the number given by .707 to arrive at true root means square voltage. Newer multimeters will give you true RMS voltage. In any event, your measured voltage should be within 5%.

There are four common types of forced air ignition systems.

1. Standing Pilot (with or without Baso switch)
2. Spark and Thermo Coupler
3. Spark and flame sensor
4. Hot Surface Igniter (with or without a flame sensor)

Before we can troubleshoot residential HVAC equipment, we must know the sequence of operation. Let's look at the standing pilot type of forced air furnace. The electrical is the simplest in this type of furnace. You will see a recognizable pattern for all of the circuits. The first circuit will be the transformer (power), thermostat (switch), gas valve (load), and common. When the switch closes (thermostat), it allows power to flow to the gas valve (load), then through the gas valve (load), then back to the transformer through the common. This is a practical application of our universal theme of **Power, Switch, Load, Common.**

However, in order for our gas valve to open, a standing pilot must be established. This too is another series circuit. A thermocouple is two wires where one wire is inside the other. When two dissimilar metals are arranged side by side and are heated, an electrical voltage is created. It is a very small voltage measured in millivolts. So the circuit in this case is; the inside wire is heated by a flame (switch) causing a millivoltage (power), this millivoltage is transferred to the pilot valve (load), and then is returned to the gas valve through the outside wire (common). This circuit maintains the standing pilot. Any break in this sequence will cause the pilot valve to close shutting off the gas supply to the standing pilot, which in turn extinguishes the flame and ends the voltage caused by the standing pilot. The standing pilot is intended to run continuously not turning on and off. The standing pilot is a safety feature in this type of furnace. The standing pilot must be intact and fully functioning for the main gas valve to open.

With the standing pilot functioning properly, the gas valve and blower motor can now be activated. The thermostat (**switch**) will close allowing power from a 24 volt supply (**power**) allowing the main valve on the gas valve (**Load**) to open, then returns back to the supply through the **common**. The opening of the main gas valve will cause gas to ignite in the burners.

The blower motor circuit follows the same circle. The 120 volts supplied to the furnace is used operate the blower motor. There are many thermal switches designed to close with a rise in temperature, but let's use a type called a snap disc switch. Snap disc switches are commonly used on mobile home furnaces. The switch closes as the burner heats the heat exchanger. This rising temperature closes the snap disc (**switch**) to close allowing the 120 volts (**power**) to reach the blower motor (**load**), power goes through the blower motor and then returns through the **common**. See figure 1-5.

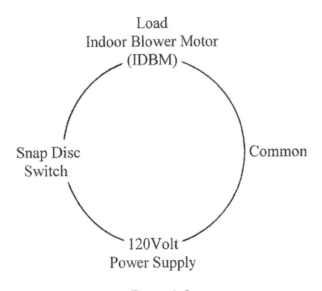

Figure 1-5

Figure 1-6 shows at all three circuits and their sequences. The three individual circuits are; pilot, gas valve, and blower motor.

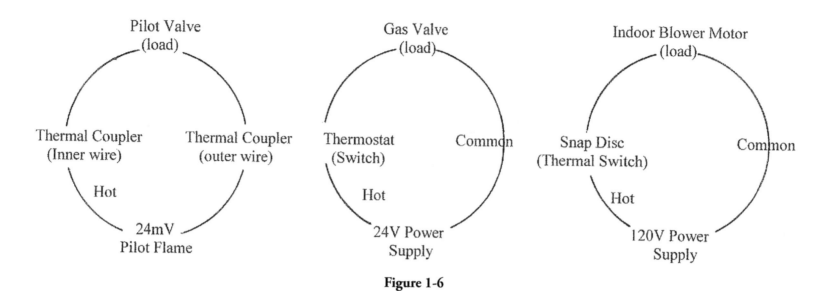

Figure 1-6

As you can see, each circuit is made up of the same components of power, switch, load, and common. You can see each circuit has a different voltage. Because each common has independent sources, means there are no parallel circuits. All of the examples in figure 1-6 are series circuits.

The difference between a series circuit and a parallel circuit is, a series circuit is independent but shares the same voltage or power supply with another series circuit. When several series circuits share the same power supply, it is considered to be in parallel. Or as stated, each series circuit is in parallel with the same power supply as other loads. Each load must be independent in a series circuit or voltage will drop below the supply voltage required for each load depending on the resistance in the load. This will be explained later in the book. The purpose of this book is to show how all circuits are in the same circle and in a simplistic solution for the purpose of troubleshooting each series circuit. In the remainder of this book, the circular figures showing circuits will now be shown as straight line circuit figures. These straight line figures are a closer representation of an actual wiring diagram. Voltage will still flow in the same directions and the basic components will remain the same.

Figure 1-7 shows a series parallel circuit.

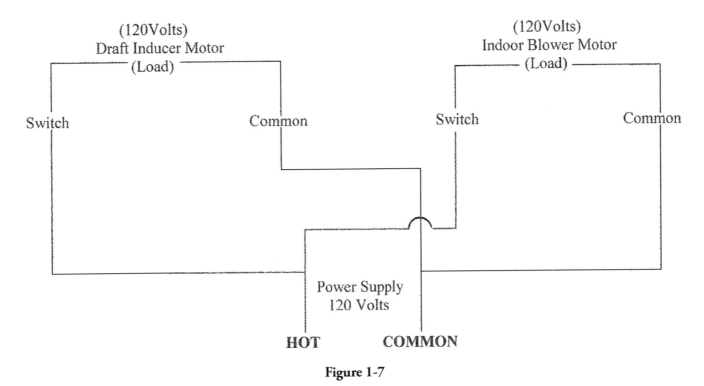

Figure 1-7

Because both loads require the same voltage, both series circuits can share the same power supply and common. The only thing that changed is the switches that operate the loads. Configured this way, they can operate independently from each other. When we apply the same source of voltage to each series, they are considered in parallel.

HVAC ELECTRICAL FOR IDIOTS | 13

If two loads are put in the same series circuit, then the voltage will be divided according to their resistance values. Figure 1-8 will show an example.

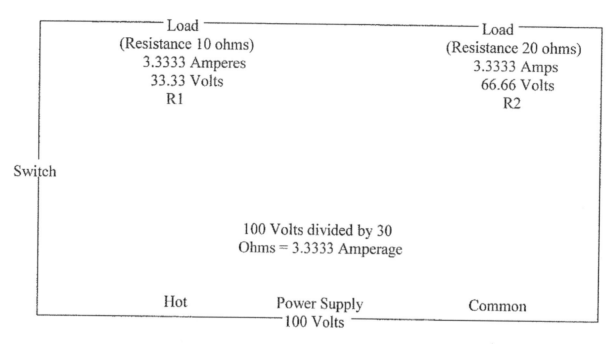

Figure 1-8

As you can see in figure 1-8, both loads require 100 volts to operate. Because both loads are in series with each other, the voltage is divided by the amount of resistance in each load.

Ohms law states that the total; voltage divided by the total resistance will be an amperage of 3.3333. Ohms law states amperage multiplied by the resistance will equal the voltage for each law. Kirchoff's second law states that the individual voltage of each load divided by the resistance is the amperage for each load which is 3.3333. The voltage for each load will equal the total voltage in the series circuit.

Where the resistance or R1 or the first load multiplied by the current 3.3333 or (10) ohms multiplied by 3.3333amps equals 33.333ampheres. R2 (20 ohms) multiplied be 3.3333equals 66.666volts. 66.666 volts divided by R2 (20 ohms) is 19.9982 amperes

This can be used in electric circuit boards; however, you will never need to learn this as we are building simple circuits and not circuit boards.

Figure 1-9 shows the same power supply in two single series circuits that are in parallel with each other. This will give you a better understanding of why two loads will never be in the same circuit.

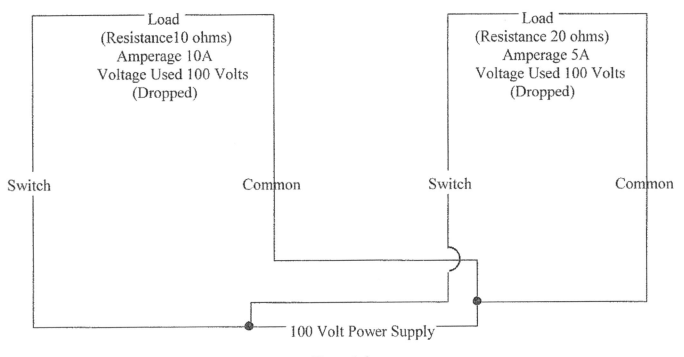

Figure 1-9

As you can see in figure 1-9, when in parallel with each other, each circuit receives the full voltage. Otherwise sharing the voltage with another load will raise the amperage too high and cause the loads to burn up and fail as in figure 1-8.

For the remainder of this book, our examples will focus on single series circuits, in parallel with their corresponding voltages. You can only parallel like voltages. As an example; you can parallel 24 volt relays, contactors, gas valves, and other 24 volt loads. The same common can be used for all 24 volt loads in parallel. In another example; 120 volt indoor blower motors (IDBM), draft inducer motors, transformers, hot surface igniters (HSI), or other 120 volt loads. The same common can be used on all 120 volt parallel circuits.

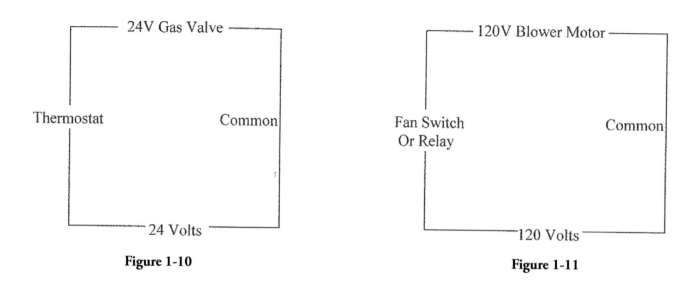

Figure 1-10 **Figure 1-11**

Figures 1-10 and 1-11 shows two separate circuits. Each requires a different voltage. Figure 1-10 is a 24 volt circuit. You cannot use 120 volts for that circuit. Figure 1-11 is a 120 volt circuit. You cannot use 24 volts for that circuit. **Using the wrong voltage can cause a fire hazard or cause serious personal injury**. Always check for proper voltage before proceeding.

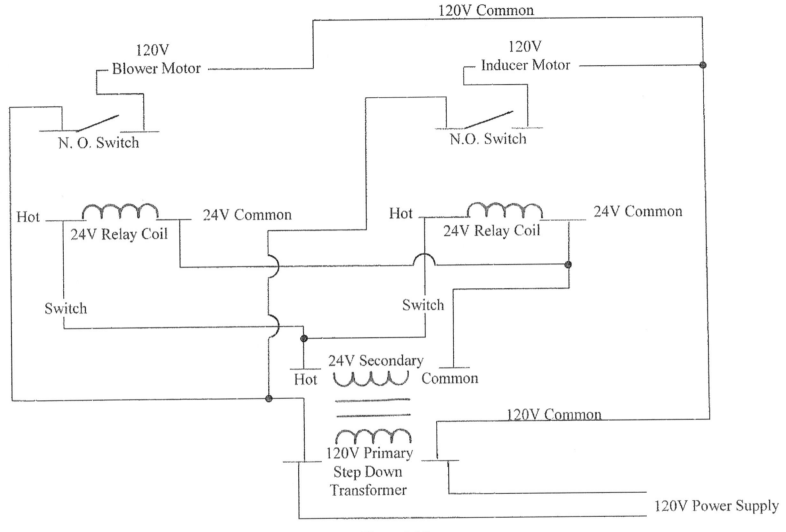

Figure 1-12

The 24 volt relay coils shown in figure 1-12 create a magnet that pulls the switch closed.
This is the symbol for a magnetic coil. ⌒⌒⌒ This coil is an inductive load and is measured in amperes.
This is the symbol for an electric heating element. ⌇⌇⌇ This heating element is a resistive load and is measured in wattage.
All loads that create an electromagnet are inductive loads. All loads that create heat or light are resistive loads.

HVAC ELECTRICAL FOR IDIOTS | 17

CHAPTER 2

Relays are use to allow a higher voltage to be controlled by a lower voltage. You will need to know how basic relays operate. A relay has a low voltage coil that will create an electromagnet when the electricity flows through the coil. This is called electromagnetism. The electromagnet is made up of a coil of wire wrapped many times around laminated metal, commonly iron. This magnet is activated by an action caused by the thermostat or by a circuit board allowing the flow of electricity. The magnet that is created by the coil will change the position of a switch to being either open or closed. When a switch is open before power is applied, it is considered ***normally open***, commonly identified as **NO**. When a switch is closed before power is applied, it is considered to be ***normally closed,*** commonly identified as **NC**. Figure 2-1 shows a basic general purpose relay.

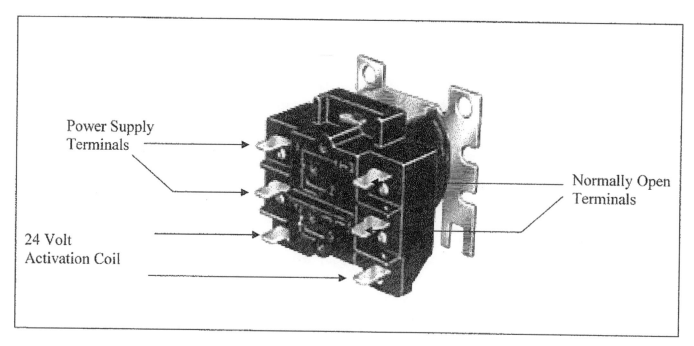

Figure 2-1

The top two terminals are normally open and are used for higher voltages. The terminals on the far left, is where power is applied. The right side is for the hot load. Terminals in the middle will be normally closed (NC). When power is applied, the switch will close between the hot and right terminals and open between the hot and center terminals. The applied power is usually 24 volts or 120 volts.

CHAPTER 3

Whether in a circuit board or separate hard wired relay, the same four items remain constant and in this specific order: **Power, Switch Load, and Common.** Once we understand this concept, troubleshooting any electrical circuit is simple. You can have or use as many switches as you want as long as they are placed **BEFORE** the load and do not require electricity. If a switch, such as a relay, requires electricity, then each relay will have its own series circuit to operate and can have only one switch. (This will be demonstrated later in this book.) In figure 3-1, consider all of the switches to be normally closed (NC) and will open only when a desired temperature is reached. Snap disc switches, limit switches, roll out switches, and other temperature type switches are examples of these desired temperature switches.

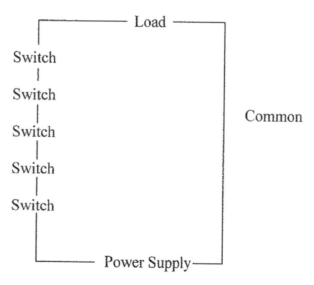

Figure 3-1

It is easy to identify the difference between hot and common. If it leaves a power supply and goes to a switch, it is hot. You can easily test this using your voltmeter. By touching one lead of the voltmeter to the wire and the other lead to ground (chassis or bare metal on the cabinet) and the voltmeter indicates a voltage, then it is hot.

Knowing how to use a voltmeter is vital to troubleshooting. You will need to know how to measure Voltage, Resistance, Continuity, Capacitance, Micro-Amperage, and Milli-Volts. It is also a bonus if your voltmeter can measure temperature.

CHAPTER 4

You will need to know the following symbols on your meter.

1. ∿ V Sine Wave A/C Voltage (Alternating Current) No polarity but a direction of flow. Preferably Self Ranging

2. ⎓ V D/C Voltage Direct Voltage Has Polarity (Positive, Negative)

3. ⊣⊢ Capacitance

4. 🔊 Continuity

5. ∿ A A/C Amperage (Alternating Current) Clamp On for whole Amps, or in series for micro Amps.

6. ⎓ A D/C Amperage (Direct Current)

7. Ω Resistance This gives you a Resistance Value in Ohms The horseshoe looking item is the Greek Omega sign.

    When reading continuity, always test your meter by touching the two wire leads together. It should read something other than **OL** (out of limit). If it reads anything other than **OL**, it means there is continuity.

    When reading voltages, make sure the decimal point is in the correct place. For example; 2.4 volts or .24 volts; but not 24. volts. The voltage should always read at least the minimum voltage. 24 volt and 27 volt readings are acceptable. A 22 volt reading is not enough voltage. The 120 volt supply should read between 110 volts and 122 volts. If the voltage is lower than the unit requires, then check by measuring the amperes range on a couple of loads that require the voltage. If more than one item draws excessive voltage, then a bad connection might be present and you will need to call to the utility supplier. There might be a loose lug in the breaker box or supply.

Let's do some troubleshooting. Figure 4-1 shows a simple series circuit. Using the sequence of operation, what part was to operate next? For example; if the burners lit and the blower motor did not come on and it was next in the sequence, we would go right to the blower motor to test for power. We follow the motor leads back to where they are mounted and test for power. This would be located after the switch but before the motor and common. If this is a multi-speed motor using a circuit board, you would test at the heat terminal and its common on the circuit board. **In order to prove whether a part is bad or not, we ALYAYS start with the part that is not working.**

We need to test the nonworking part for power first. If it is not getting power, we cannot expect it to work. You would not consider a box fan in your home to be nonworking until you plugged it in. We ALWAYS go to the nonworking part and check for voltage. If it has voltage and does not work, then the part is bad, in this case the motor.

If there is no power to the nonworking part, then we need to check the power source. If the part doesn't have power but everything to that point is working correctly and we are testing the circuit board, then the circuit board is bad. The test lead for the common can be placed anywhere from the common on the motor to the common on the supply. You can also place your DMM common at any established common from other parts, provided it is the same voltage.

To save time on furnaces with circuit boards, look for an LED light, if it has one. If it is lit, both 24 volts and 120 volts are present and there is no need to check fuses, breakers, or the step down transformer.

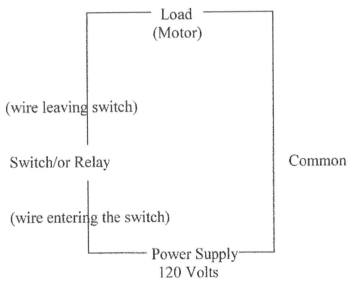

Figure 4-1

UNDERSTANDING THERMOSTATS

Before we go further in troubleshooting, you need to familiarize yourself with thermostats. There are several types: Mercury, magnetic, Milli-volt, Light and digital are examples. We also have single stage and two stage heating and/or cooling. The most common thermostats are magnetic (commonly used in mobile homes) and digital and mercury thermostats (used in residential homes). We will use digital thermostats for our examples. Testing will be the same for each type. All of these thermostats are mounted on a sub-base. That base is mounted on the wall and the wires to the furnace or AC will be attached to that base. The thermostat is then attached to the base making contact to pre-assigned locations on the base. If it is a heat only thermostat, it will have only two wires (red and white). Thermostats have a heat anticipator which maintains a two degree difference in cycle time based on the amperage drawn from the gas valve. This has to be adjusted on mercury style thermostats only based on amperage measured from the gas valve. Steps for setting the heat anticipator are not covered in this section. If the thermostat is being used for heat, A/C, and fan only, then you will have four or six wires going to the thermostat base. Four wire bases use red, white, green, and yellow. Six wire bases will additionally have brown and blue wires. *If you are installing a new system, use the six wire bundle for future use.* These wires go from the thermostat to the furnace.

Red is used for the 24V power from the transformer to the thermostat and we use a mode (heat, A/C, or fan only) for the power to come out of the thermostat. Green is for fan only. White is for heat mode. Yellow and blue can be used for A/C. Power-Robbing thermostats will require the brown wire connected to the common (C) terminal, which is also the transformer common. In the A/C mode, the power leaves the thermostat using the yellow or blue wire (whichever is being used for A/C on the Y terminal between the sub base and furnace. The green wire is used for the fan and is interlocked in the thermostat. This means when the A/C or Y terminal is used to turn on the A/C the G or fan terminal is on at the same time as the Y. The A/C compressor and the fan must come on at the same time.

When the fan is in the A/C or fan only mode it will operate at the highest speed on the Indoor Blower Motor (IDBM). In heating mode, we use the lowest speed possible. This is done to move warm air without cooling it. This is determined by the heat rise established by the manufacturer and is located on the furnace data plate. Heat rise will not be covered in this book.

For the sake of keeping this book simple, we will be using the yellow on our thermostat to operate A/C in the A/C mode.

UNDERSTANDING ELECTRICITY

Before we start the troubleshooting procedure, let's take a look at electricity. This section does not need to be understood to apply the troubleshooting techniques in this book, however, it does help with understanding what you will be looking for and what you will be measuring. All things (masses) are created from atoms. We cannot actually see atoms but we can understand their structure and use. Atoms are made from positively charged particles (protons) and non-charged charged particles (neutrons) that make up the nucleus. Negatively charged particles (electrons) orbit around the nucleus. It is the electrons that we will be measuring.

Electromotive force is the pressure applied to the electrons to move them. The amount of pressure (electromotive force) placed on electrons to get them to move is measured in *voltage*. The amount of those electrons that we move are measured in *amperage* and is commonly called the *current*. Listed below are the most common formulas used to calculate the amounts of each in a series or parallel circuit.

You will need to know how to use both **Ohm's law** and the **Power Law**. The next page will put these in a better form to understand. You will need to know the scientific/engineering symbols for each unit of measure.

- The capital letter **E** is used to indicate the electromotive force and is measured in *voltage* and can be expressed as capital letter **V**. An example: 24V. When using this symbol we must indicate the type of voltage being used; alternating current or direct current. An example: 24 VDC or 120VAC.
- The capital letter **I** is used to indicate current. Current is measured in *amperage*. The letter A is used after the number with a space between. An example would be: **I = 3 A.**
- The capital letter **R** is used to indicate *resistance*. This is the measure is resistance against the applied pressure used to move through a load expressed by the Greek Omega symbol Ω.
- The capital letter **P** is used to indicate *power* and is measured in *watts*.

There are two types of loads, **Inductive** and **Resistive**. An inductive load would be something that crates an electromagnetic coil or field and is measured in *amperes*. Examples would be: 24 V relays, 24 V gas valves, 120 V electric motors, and etc. Resistive loads give off heat and/or light and are measured in *watts*. Examples would be: light bulbs, electric stove burners and hot surface igniters (HSI). We can now use these symbols for use on the next page.

When solving for one or the other, place a multiplication symbol between the **I** and **R** signs and use the lines between the top and bottom as division. Cover the letter you are looking for in the figure 4-2 and the diagram will tell you how to solve for that item.

- E= I x R or Electromotive Force (voltage) equals Current (Amperes) multiplied by the Resistance (ohms).
- E / I – R or Electromotive Force (Voltage) divided by Current (Amperes) equals Resistance (Ohms).
- E / R = I or Electromotive Force (Voltage) divided by Resistance (Ohms) equals Current (Amperes).

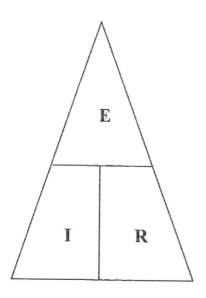
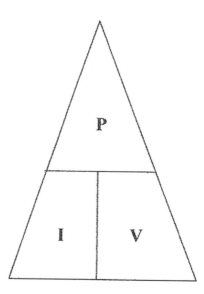

Figure 4-2

Now that we have looked at the basics, let's build a few simple circuits. Let's look at the minimal parts needed to a completed, controlled circuit. Start with just a wire and lay it out in a circle. **If we always think of this circle, then we can troubleshoot a circuit accurately every time.** It is important to keep **Power, Switch, Load,** and **Common (PSLC)** always in mind. **It is important that these four items must be committed to memory in that exact order.** Efficient troubleshooting is based on knowing these four items and using them in the specific order given. Deviating from this prescribed order will make troubleshooting difficult or impossible.

Using a conductor wire lay it out in a circle. Because there is no power supply, switch, load, or common; we do not have a circuit. We simply have a wire in a circle. If you cut the wire at the bottom, we will now have two distinctive ends. If we add a power supply to these two ends, we now have the beginnings of a circuit. See figure 4-3. For training purposes we will assume that the flow of power is clockwise. In an actual circuit the direction of flow can be established by which side the switch is on. A switch will NEVER be located on the common side.

DANGER! NEVER apply a power supply to a conductor that does not have a load! A power supply without a load is a short and can cause extreme damage to person and or property!

Figure 4-4 shows where a load has been added to the circuit. Let's assume we are building a 24 V circuit using an inductive load creating an electromagnet such as a relay or operating a gas valve.

Figure 4-5 shows where a switch has been added. Because of the location of the switch, we know that the flow of electricity is clockwise. The switch is added on the hot side only. Figure 4-5 shows a completed, controlled circuit: **Power** Supply, leading to a **switch**, leading to a **load**, and completing the circuit on the **common**, back to the power supply.

Figure 4-3 and 4-4 and have been shown for educational purposes only to help the learner understand the steps in building a circuit. NEVER build a circuit as shown in figures 4-3 or 4-4. Severe injury to person, equipment, and property can result!

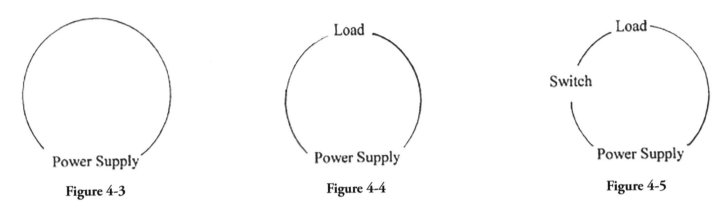

Figure 4-3

Figure 4-4

Figure 4-5

We can add as many switches as needed on the hot side of the circuit (between the power supply and the load). The two most used symbols are used to indicate a temperature switch (figure 4-6) or a pressure switch (4-7).

Temperature Switch

Figure 4-6

Pressure Switch

Figure 4-7

Switches can be combined in useful arrangements. Negative Pressure Switches (Vacuum Switch) can be added to the circuit with Temperature Switches (Roll Out Switches or Limit Switches) as long as they are rated for the given load, voltage, and amperage; and do not require voltage to operate, as in a relay or contactor. See figure 4-7.

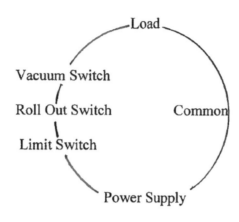

Figure 4-7

We must start our troubleshooting at the faulty load. The sequence of operation will always determine what the next load is that will be operated. Consider the following sequence; a draft inducer, a vacuum switch and a hot surface igniter. The sequence is that the draft inducer causes the vacuum switch to close thus allowing the hot surface igniter to heat. Our examination indicates that the hot surface igniter does not heat. **We will start our troubleshooting at the hot surface igniter.** Beginning our search anywhere else will be a waste of time.

We begin by testing the part or load that is not working and work backwards from there until we find the break in the sequence of operation. **BEGIN AT THE FAULTY PART.** *(All of our examples will be using 24 V unless otherwise specified.)* Is it receiving voltage? We cannot say a part is bad if it has no power to operate. It would be like saying your T.V. is broken when it is not plugged in. If we have voltage on both sides of the load and it does not operate, **the part is bad**.

If the suspect part or load does not have power, then we must first establish which side is bad, the entering side or the exiting side. Also, we need to know the required voltage is required to operate the part or load. Is it 24 V, 120 V, 240 V, 208 V, 227 V, etc? **Always verify the required voltage.**

Once the voltage is verified, use your meter in AC Volts to test. The x in figure 4-8 indicates where to place the test leads. Touch one lead to a terminal on the part or load marked x and ground the other lead to the chassis. If it shows the correct voltage that is the hot side, which means that the other side is the common side. Please study figure 4-8 carefully and 4-9 carefully.

Do the same with the other lead from the load marked x to the chassis ground as shown in figure 4-9. Whichever shows voltage, that will be the hot side. If no voltage is found in either figure 4-8 or 4-9, then power has been lost. If we have voltage at both load terminals, then we have lost the common.

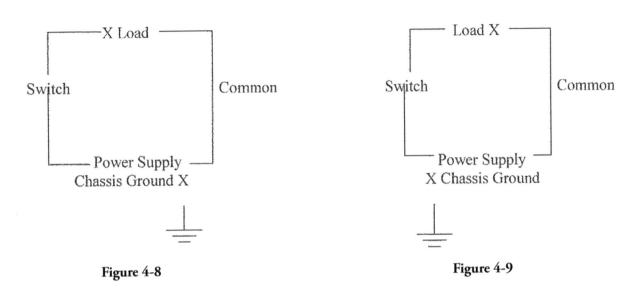

Figure 4-8   **Figure 4-9**

If you have voltage at both figures 4-8 and 4-9, then you must establish which is the common. Determine what side comes to the switch from the power source. The opposite side will be the common. Using your digital multi-meter (DMM) in continuity mode, verify that you have continuity from that terminal to the cassis ground. If we find that the other side that does not have the switch (the common side) has power, we will need to re-establish the common by attaching to the chassis or other known common. The other side will be the hot side.

If a circuit does not show power on either side of the load, check the transformer for voltage at the source. If voltage is found at the source, then both the hot and common has been lost somewhere in the circuit. **Always re-establish the common side first.** There are no switches on the common side and it is easily repaired. Rather than re-wiring the entire common side all the way back to the power, you can ground the common to the chassis. This can also be accomplished by connecting the common on the inoperable part to a known common on a working part. Now that we have a common established, we can use it to find where the hot side has stopped. The hot side will be between the power supply and the load.

We will now have put one of the leads to common and the other side will start at the hot side of the inoperable part. We will continue moving and place the hot side lead backwards toward the power supply until voltage is found. If switches are involved in the hot side, each will need to be tested going backwards from the source part to the power supply.

- If power has stopped between the load and the last switch we tested, then the wire between them is bad and needs to be replaced.

- If you have power coming into the switch but not coming out of the switch, then the switch is bad and needs to be replaced.
- If you have power on both sides of a switch but no power to the next switch, then the wire between the switches is bad and needs to be replaced.
- If there is power in and out of the second switch, but no power is found at the next switch, then the wire to the next switch is bad and needs to be replaced.

Every time you make a repair, you must start over from the common to the load and work your way back until you find voltage. You will do this until you find where the voltage reaches the load. **It is important to know if a switch is normally open or normally closed. For example a vacuum switch is normally open and will not close to allow current to flow until the draft inducer motor is operating thus pulling the switch closed. If an indoor blower motor (IDBM) fails, a high temperature safety limit switch will open, and stops the gas valve from opening due to the high temperature. CAUTION: Never leave a jumper wire on a safety! Fire or destruction to property or person may occur.**

Any part that requires electricity to operate is based on this same circle principle. If they share voltages, it is called a parallel circuit.

It is helpful if you have a jumper wire with alligator clips attached to each end. There is always more than one way to verify your findings. On switches and only switches, you can place a jumper wire on both sides of the switch to verify if the switch is at fault. You must always turn off power when jumping across the terminals of a switch. Another way of testing a switch is to remove the wires from the switch and check the switch for continuity. To do this, put your DMM on continuity or ohms. When you test the switch it should change from OL (out of limit) to any number. This will mean the switch is closed or on.

A switch can be tested with the power on. Put your DMM on AC voltage and place the leads where the two wires are attached to the switch. If you read the full voltage on the DMM, the switch is open. If it reads 0 volts, the switch is closed. **You cannot jump out a load. This is very dangerous and can cause serious damage to your health.**

Chapter 5 shows how to trace back from an inoperable load.

CHAPTER 5

The examples in chapter 5 are intended to be used as a study for the step by step procedure when troubleshooting a fault. In the examples the X is where you are to put your digital multi-meter (DMM) leads. In the example on this page, we are exploring the possibility of a broken switch.

You will use this procedure every time you troubleshoot an electrical circuit. Figure 5-1 shows where you begin your testing. Make sure your DMM is in Volts AC. Since the power supply is 24 Volts, you should get a reading of 24 Volts across the switch. Caution: make sure to observe where the decimal point is. .24 Volts or 2.4 Volts is NOT 24 Volts. If the test shows that we have the correct Voltage and the part does not work or function, proceed to the next step shown in figure 5-2.

We cannot say the part is bad if there is no Voltage. Figure 5-2 indicates where your next measurement is to be done. We are establishing if the problem is before or after the load. If testing shows we have no Voltage then, test as shown in figure 3-5. If testing as indicated in 3-5, shows we do not have Voltage, move on to testing as shown in 5-4. If Voltage is still not found then, test as shown in figure 5-5.

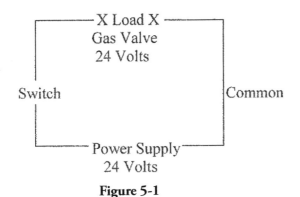

```
        ─ X Load X ─                              ─ Load X ─
         Gas Valve                                Gas Valve
          24 Volts                                 24 Volts

 Switch           │Common            Switch              │Common

        ─Power Supply─                           ─Power Supply X─
          24 Volts                                 24 Volts
```

Figure 5-1   **Figure 5-2**

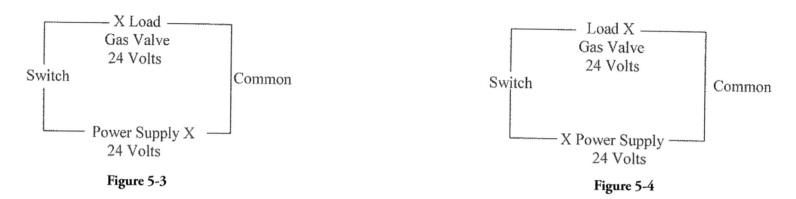

Figure 5-3

Figure 5-4

Figure 5-5 shows the next place to test. After testing as indicated we now have Voltage. We cannot have a reading from the same side, so we now we need to test the common side of the power supply. Now we discover that we have no Voltage. **ALWAYS make sure the common you use to find where you lost the hot side is the same voltage.** In other words, do not use a 120 Volt common to find a 24V hot or vise versa. If there is no Voltage found in figure 5-6, keep one lead on the common side of the power supply and move the other lead to the opposite side on the load as shown in figure 5-7. If we still have no Voltage, move the test leads as indicated in figure 5-8.

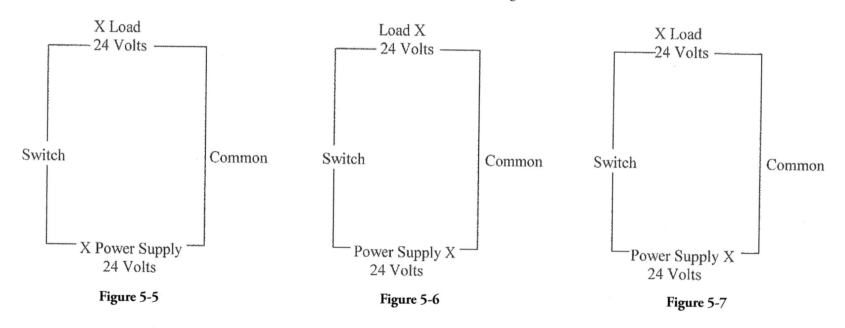

Figure 5-5

Figure 5-6

Figure 5-7

In figures 5-4 and 5-5 our Voltage has been found on your DMM. The side of the power supply **ALWAYS** determines the side that is bad. In the examples prior, it was our hot side. The reason Voltage was found on both sides of the coil is because the coil is still intact and Voltage was feeding through the coil. However, when we are able to measure Voltage on the hot side of the load and the hot side of the power supply at the same time means that this is the side that is open. This is because we are supplying the meter with the side that is missing from the power supply.

Figure 5-8 shows the next test site for our leads. You will now be testing the common side of the power supply to trace back to where power was lost. Keep the lead on the established common side of the power supply and place the other lead on the switch terminal traced back from the power supply as shown in figure 5-8. If we still have no power, the next measurement will be from the common side of the power supply and the terminal on the switch from the hot side of the power supply as shown in 5-9. If we now 24 Volts is found, the problem is a bad switch. The switch is open and does not allow 24 Volts to travel through the switch.

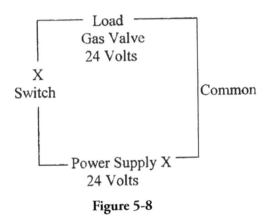

Figure 5-8

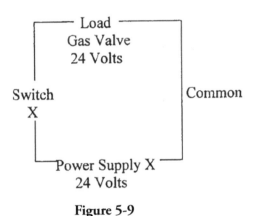

Figure 5-9

Let's look at another example. In this case the problem will be a bad common wire.

As always, we start at the load. In this case it will be a 24 Volt gas valve. Keep in mind that it does not matter which side of the power supply where you start. In this example, we will start on the common side.

In figure 5-10 Voltage is not found. This finding indicates that one side of the circuit is missing.

In figure 5-11 we find Voltage. We started on the common side of the power supply. The fact that we used common to start our testing was the side we provided to our DMM. The reason it showed on that side of the gas valve is because the 24 Volt coil in the gas valve is intact allowing the hot side Voltage to go through the gas valve and is shown on the DMM.

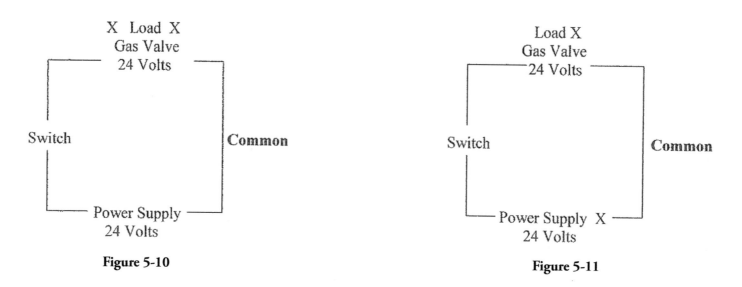

Figure 5-10

Figure 5-11

Let's look at another example. In this case the wire from the switch to the load is broken. Please be reminded, the X is where the leads are to be placed.

In figures 5-12 through 5-15 no Voltage is found.

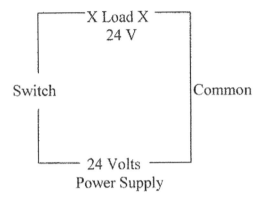

Figure 5-12

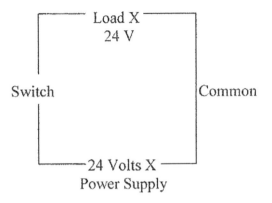

Figure 5-13

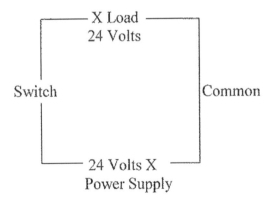

Figure 5-14

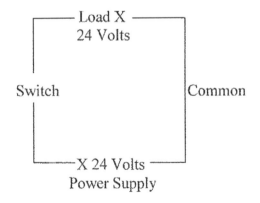

Figure 5-15

The next test will be from the hot side to the power supply. (See figure 5-16) When testing as in figure 5-16, Voltage is found. Because Voltage is found on the same sides again, the leads need to be switched to test common of the power supply and the lead to the common side of the load as in figure 5-17. If no Voltage is found, the common side of the power supply is used to find where Voltage is lost. When tested as in figure 5-18 Voltage is not found. We then test moving towards the power supply by placing leads on the common side of the power supply and to the end of the wire leaving the switch to the load where Voltage is found. This indicates that power made it through the switch but not to the 24 Volt load. (See figure 5-19)

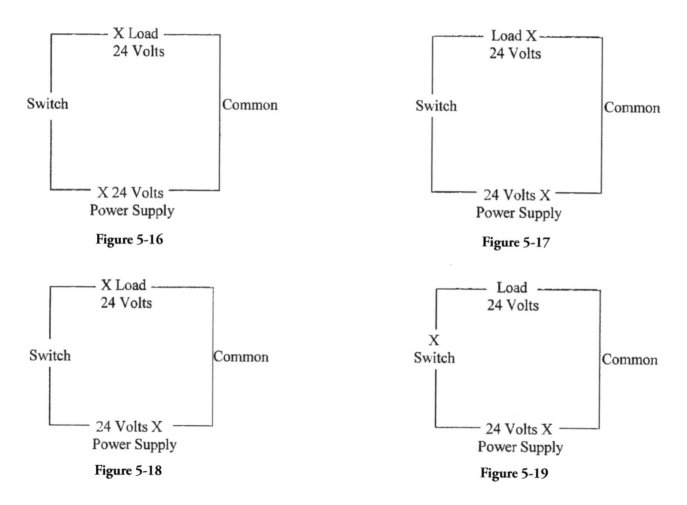

Figure 5-16

Figure 5-17

Figure 5-18

Figure 5-19

The following pages show more examples.

Example 5-1 will provide you the opportunity to complete the diagnostic procedure on your own. The next few examples will show the testing procedure to be used below the figures. Please refer to those procedures as needed. Remember we always start our test procedure at the load. Assume the switch #3 is bad.

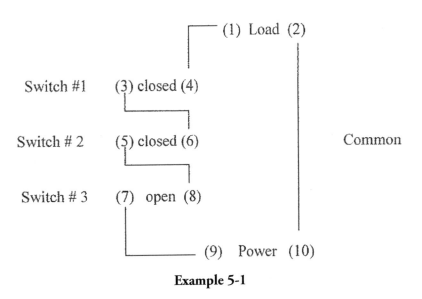

Example 5-1

Test at the load:	1 and 2
Then	10 and 1
Then	10 and 2
Then	9 and 1, Voltage is found
Then	9 and 2, Voltage is found

Using the DMM Voltage has been found at positions 9 and 1, and 9 and 2. *Because we have Voltage at 9 and 1, and they are both on the same side (hot side), we have established the problem is on the hot side (1 through 9). Remember, the power supply determines the side where the problem exists.* We need to test the common side (2 through 10) to find where the problem switch is located. In other words, where the Voltage shows up again.

Test at:	10 and 1

HVAC ELECTRICAL FOR IDIOTS | 37

Then	10 and 4
Then	10 and 3
Then	10 and 6
Then	10 and 7, where Voltage is found.
Then	10 and 5
Then	10 and 8

At 10 and 7 Voltage is found but no voltage is found past 10 and 7. This means power has made it to 10 and 7 but not past it.

Example 5-2 assumes that switch # 2 is bad.

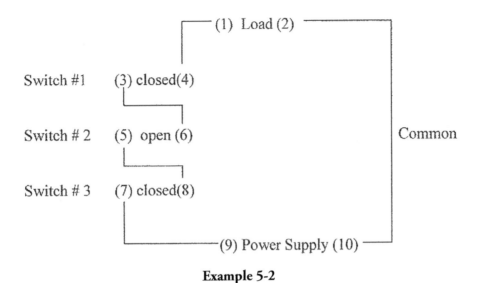

Example 5-2

Start testing at:	1 and 2
Then test	10 and 1
Then test	10 and 2
Then test	9 and 1, Voltage is found
Then test	9 and 2, Voltage is found

Voltage being found at 9 and 1, and 9 and 2 indicates the problem is on the hot side.
The common side needs to be tested to locate the problem.

Test at:	10 and 1
Then test	10 and 4
Then test	10 and 3
Then test	10 and 6

Then test 10 and 5, where we find Voltage.

Since Voltage stopped at 5, switch #2 is either bad or open. If we had Voltage at 8 and not at 5, then the wire between 8 and 5 could have been bad. The procedure requires that the path must follow the flow of current or Voltage. We are not allowed to randomly test by jumping around.

Example 5-3 assumes the problem is in the common.

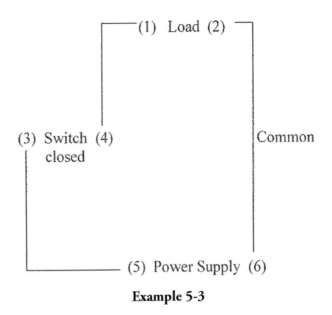

Example 5-3

First test:	1 and 2
Then test	6 and 1
Then test	6 and 2

If 6 and 2 had been tested first and shown a Voltage, then it would be necessary to test 6 and 1. If Voltage had been shown at 6 and 2, then that side is the bad side of the circuit. If that were the case, you would have to use the hot side of the power supply to find the bad part or where the Voltage stopped. If testing from 5 and 6 indicated a Voltage and the common was good, then Voltage should show from 6 and 2. If not, there is no common to the load.

Example 5-4 assumes the load is bad.

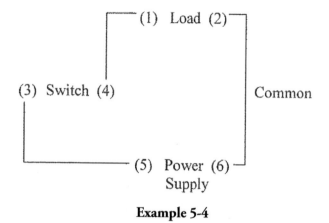

Example 5-4

First test at 1 and 2 and Voltage is found but the load does not operate. This means the load is bad.

Example 5-5 assumes the power supply is bad.

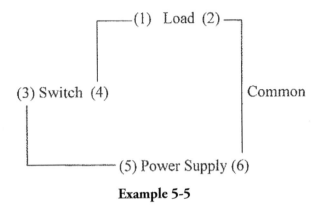

Example 5-5

Test at	1 and 2
Then test	5 and 1
Then test	5 and 2
Then test	6 and 1
Then test	6 and 2
Then test	5 and 6 If you have no Voltage, you can check for Voltage anytime before or after this test.

Example 5-6 assumes the wire between switch #1 and switch # 2 is bad.

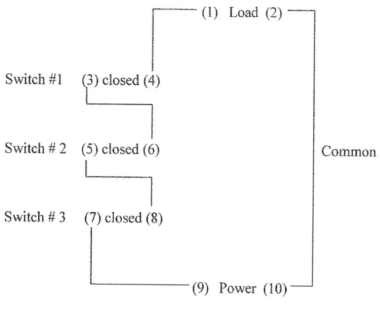

Example 5-6

First test:	1 and 2
Then test	10 and 1
Then test	10 and 2
Then test	9 and 2 or 9 and 1 and Voltage is found.

This means the hot side has the problem. Now the common side of the power supply needs to be tested.

Test	10 and 1, Voltage is not found.
Test	10 and 4, Voltage is not found
Test	10 and3, Voltage is not found
Test	10 and 6, Voltage is found. This indicates that there is a Voltage disconnect between 6 and 3.

CHAPTER 6

Capacitors

CAUTION: Turn OFF power and discharge the capacitor by shorting out the terminals or use the manufacturer recommended discharge method. The wires must be removed from the capacitor before testing. Testing capacitors is relatively easy. There are three basic types of capacitors in HVAC service. They are **start**, **run**, and **duel** capacitors. Duel capacitors combine start and run. On the duel capacitor the smaller number refers to the run portion and the larger number refers to the start portion of the capacitor. A small u and a small f are present after the number and stands for Micro Farads. Farads are the unit of charge.

Run capacitors are typically reserved electricity that is used for motors when larger loads are drained from a motor's load. It provides a constant electrical flow to the motor. All loads will have an increase in Amperes when voltage is taken away from a load. This will cause overheating and breakdown of insulation in wiring coils in motors and can cause shorts, opens, or lead to premature motor failure. The run capacitor always stays in the circuit and prevents a voltage drop.

Start capacitors give an instant boost of electricity all at once to help the motor overcome inertia. Once inertia is overcome, the start capacitor drops out of the circuit and the run capacitor takes over. Capacitors have a plus or minus of a given percentage range where they can still be considered good.

Our DMM should be able to read capacitance. The symbol on the meter should be uf. EXAMPLE: Assume the run capacitor reads 5 uf, + / - 6%. This means 5.3 uf is the highest allowable reading and 4.7 uf would be the lowest allowable reading. If the capacitor does not fall within this range, it needs replacement.

Let's look at a duel capacitor. There will be two numbers. One number for the start capacitor and another number for the run capacitor. EXAMPLE: 25 + 5 uf, + / - 5%. This means our start capacitor is 25 uf and the run capacitor is 5 uf. There will be three terminals. One will say **FAN**, another will say **HERM**, and the third will say **C** (common). The **C** (common) will be used for measuring each terminal. Set your DMM to capacitance. Place the meter leads to **C** and **FAN**. This will be the smaller number, 5 uf. Add or subtract the 5% from this number. Next, measure between **C** and **HERM**. This is for the compressor and will be the higher of the two numbers, 25 uf. Add or subtract the 5% from this number. If either side does not fall within this tolerance, then the start/run capacitor needs replacement.

Another test that can be performed is a short test. Set your DMM for continuity. Touch one lead to one of the terminals and the other lead to the side of the capacitor. Do this to all terminals, one at a time. There should be no continuity reading.

Capacitors can be used in conjunction with each other. If you need a 20 uf capacitor and only have two 10 uf capacitors, you can use them by wiring them in parallel as shown in figure 6-1. Using this configuration can also help if the 20 uf capacitor consistently over heats by spreading the heat distribution over two capacitors. ***NOTE: AC Voltages do not have polarity, so terminal use does not matter.***

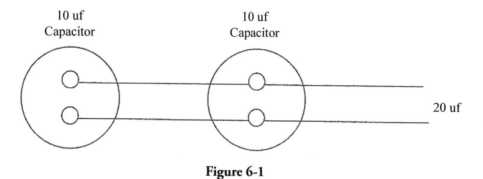

Figure 6-1

The opposite is equally possible. If you need a 10 uf capacitor and you have only 20 uf capacitors. They can be wired in series to make a 10 uf capacitance output as shown in figure 6-2.

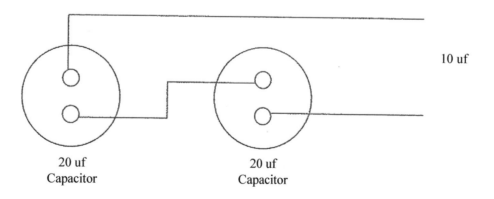

Figure 6-2

CHAPTER 7

Marketing Tips

Many money making opportunities are overlooked when servicing HVAC equipment. Preventative maintenance should be used instead of reactive maintenance. Offer your customers a preventative maintenance package that includes cleaning and testing the most vulnerable HVAC items.

The most overlooked maintenance items are: Capacitors, Motors, Contactors, and Relays.

Capacitors are commonly overlooked. If they show signs of swelling, replace it even if it works. It might not be within the capacitance range causing a higher amperage draw on the motor. Test the Amperage draw on the motor and compare that result with the amperage draw listed on the Data Plate. Just because something is running does not mean it is running well and failure could be just around the corner.

Motors can be the cause of another mistake. If you need to replace a motor, make sure the new motor is identical to the one being replaced. Often times a motor of different amperage is substituted in an effort to save money. If the amperage of the new motor is higher than the amperage of the original, the additional draw can cause circuit board failure. What you saved in the cost of the new motor will be more than offset by the cost of a new circuit board. Make sure the amperage is the same for both motors.

Contactors are often times the worst offenders. Testing the power going in and out of the contactor, should not drop more than 5%. A higher Voltage drop will cause a rise in amperage. This higher amperage is the enemy. More amperage will cause the customer more money to operate the equipment. It is the technician's responsibility to ensure efficiency without failures.

Relays can become worn and cause a voltage drop through the switch. This can cause a higher than normal draw in current causing premature failure. Measure the Voltage going in and out of the relay. If there is a voltage drop, the relay must be replaced.

Every residential furnace and air conditioner has a data plate or data sticker attached that shows the amperage draw for the whole furnace or condenser unit. If the amperage is above the allowable amount, start checking with the largest loads and work backwards until you find the load that is drawing the excessive amperage.

CHAPTER 8

Lubricating Motors and Shafts

This chapter will outline proper procedures for lubricating motors and air handler shafts. Many technicians and maintenance personnel are doing this vital service incorrectly. Proper oiling of motors will not only prolong the life of the motor, but allow it to run more efficiently. Bearings and shafts improperly lubricated will cause drag which will lead to a higher amperage draw. This in turn could cause premature failure of relays and possibly motor failure.

The proper way to lubricate a motor is make sure the bearings and shaft are cool. Lubrication will not adhere to hot bearings, shafts, or bushings.

A/C and PSC (Permanently Split Capacitance) motors are most commonly found on residential HVAC equipment. These motors typically have bushings rather than bearings. The following procedure will work on both.

Oiling: If the dust stops or caps are missing, the oilers are probably full of debris. Using WD-40 is a good cleaning agent that will not dry out the bearing. WD-40 will dissolve and flush most debris. Allow sufficient time for the WD-40 to dry. **DO NOT RUN THE MOTOR.** When the WD-40 is dry, add 3-4 drops of manufacturer recommended lubrication and hand-rotate the shaft until it is well lubricated. The oil will now remain on the shaft when the motor is started. If you do not have oil plugs for the oilers, Q-Tips can be soaked in oil and used in the oilers. The cotton tips acts as a wick for the oil and prevent debris from getting to the shaft and bearings.

Greasing: Allow shafts to cool completely before greasing. Grease will not adhere to hot metal. Use manufacturer recommended grease and do not over grease. Over greasing can cause seals to become damaged. Grease that is slung off the shaft can cause dirt and debris to accumulate on the inside windings, which can cause overheating and motor failure. Rotate the motor by hand so the grease can thoroughly coat the shaft before starting the motor.

Motor Cleaning: I recommend using dry nitrogen for cleaning a motor. Small tanks of dry nitrogen can be purchased for modest cost. A set of manifold gauges fitted with a blow gun attachment must be used. ***CAUTION: Always use a regulator when using blow gun. NEVER attach a hose directly to a pressurized tank. Serious harm or death may occur!*** This is a great method for blowing debris off any motor and does a great job of cleaning the windings. Cleaning debris off a motor is important because most motors are cooled air moving over the motor. When debris builds up, the motor can easily overheat. If you look online for motor cleaning methods, you will see many videos showing the use of cleaning with a brush. Using a brush for cleaning does not do completely remove the buildup of debris and can lead to the coils inside the motor frame to become loaded and suffocate the required air cooling movement.

CHAPTER 9

Belt Maintenance

Proper functioning of belts on blowers and air handlers is of vital importance. When checking belts for correct tension, it is important to also check the sheaves (pronounced shiv and commonly referred to as a pulley). When a belt slips, heat is created through friction. This heat is then transferred to the shaft and then into the motor bearing, causing the lubrication to burn of, causing premature bearing failure. Inspect belts for cracks or shining surfaces. If needed, replace belts with one of the proper length and width.

Inspect the sheaves for bluing or thinning. Slippage of the belt causes excessive heat and then bluing can occur on the sheaves. Sand the surface of sheaves with 80 grit sanding cloth if they are shiny, glossy, or blue. This will allow for positive grip of the belt when the motor starts.

Belt deflection should be ¾" when pressing down on the belt between pulleys. A good way to align the belt is to lay a laser pen in the groove of one pulley and aim the beam towards the other pulley. The light should line up with the center of the other pulley. The belt must be removed to do this procedure. When you re-install the belt, check to make sure it deflects ¾".

A loose or misaligned belt can cause the blower to slow down and lead to overheating of the furnace. The opposite can be equally damaging. Over-tightening of the belt will raise amperage leading to premature bearing wear and belt stretching. If the pulleys are less than three inches and have a very tight radius, use a cogged type belt to keep the belt cords and fibers from breaking.

CHAPTER 10

Testing Compressor Windings

CAUTION: Power must be off and wires removed before testing! Sometimes the wiring diagram is not available or we do not know where the wires are to be connected. The following procedure will help determine where the wires are correctly placed. It is possible to test single phase windings and determine start, run, and common terminals. Possibly you might want to know if a constantly tripping breaker is the fault of the compressor.

The three terminals on any single phase motor are Start, Run, and Common. To determine which is start, run, or common we use the DMM set to ohms. The ohms symbol will appear on your meter as Ω. To test that the DMM is operating correctly touch the leads together. The meter should respond as OL followed by some number. The better the connection between the leads, the closer to zero the meter will indicate.

It is best if you use a wire with alligator clips attached to each end. Attach one alligator clip the end of the lead and the other alligator clip to the terminals on the compressor. This procedure will give you more stable readings. Draw a picture on a sheet of paper as shown in figure 10-1. Take a measurement between terminals 1 and 2. Write this reading between the 1 and 2 terminals. In the example provided in figure 10-1 the reading indicated 1.7 ohms (Ω). Next, measure between terminals 1 and 3 using the same procedure. In the example shown in figure 10-1, the reading is 2.3 ohms (Ω). Finally, take a measurement between terminals 2 and 3. The example provided indicates 4.0 ohms (Ω).

- Because the reading between terminals 2 and 3 is the largest reading, this lets us know that those two terminals are the start and run terminals. By process of elimination, terminal 1 must be the common terminal.
- Because the reading between terminals 1 and 2 is the smallest, this lets us know that terminal 2 must be the run terminal.
- By process of elimination, terminal 3 must be the start terminal.

For the windings to be good, then common, start, and run ohm readings should be added together and that total should be within 1 to 2 ohms of the combined ohms of the start and run readings. You can also check for grounded windings by placing your DMM to continuity and placing one lead on a terminal and the other lead on the bare metal of the compressor. The meter reading should not change and continue reading **OL**.

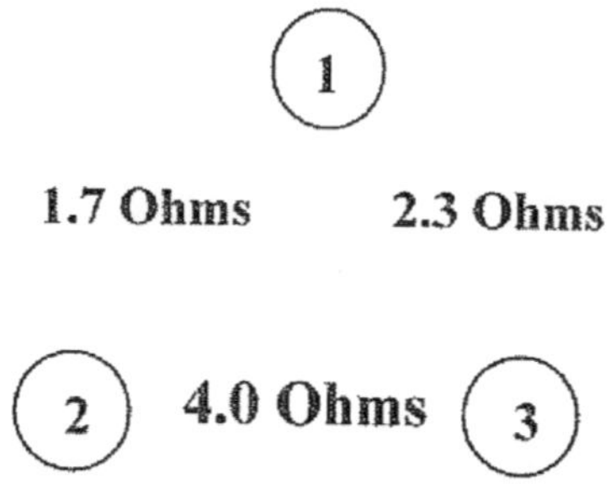

① 1

1.7 Ohms   2.3 Ohms

② 2   4.0 Ohms   ③ 3

Figure 10-1

CHAPTER 11

Blower Wheel Removal

1. Remove the blower assembly from the heating unit on furnace.
2. Lay on a flat stable surface with the blower wheel set screw facing upward.
3. Remove the set screw completely.
4. Push the blower wheel downward on the shaft. If it does not move, do not force it. Place a suitable deep wall socket that fits over the shaft and touches only the arbor. Using a wooden hammed, drive the blower wheel further down the shaft 3/8 to ½ inch.
5. Use a light 80 grit emery cloth to polish the exposed portion of the shaft.
6. Use a zoom spout oiler and lubricate the flat portion of the shaft.
7. Hold on the blower wheel by hand so it does not move.
8. Using an adjustable wrench, place it on the shaft tightly and rotate the shaft with the blower wheel not moving.
9. When the shaft is very loose from the blower wheel, turn the blower housing over and remove the mounting bolts.
10. Place the motor housing over a trash can or other suitable container that will allow the motor to fall without causing it damage.
11. Shake the housing until the motor is free from the blower wheel.
12. Replace the motor and wheel housing using the opposite procedure. Be sure to oil the motor oil tubes while they are still easily accessible. Make sure to install the motor into the housing so the oil tube ate pointed upward and the capacitor wires can easily reach their intended locations.

CHAPTER 12

Bad Gas Valve or Frozen Meter

This procedure requires the use of a manometer. Manometers can be purchased at most HVAC supply houses or online. Make sure to use one that measures in WC, (Inches of Water Column).

Remove the port plug on the manifold and insert the barbed fitting that comes with the manometer. Attach the hose of the manometer to the barbed fitting and also to the manometer. Start the furnace. Make sure the manometer is set to WC. It might be expressed as In-Wg on some meters. Not all gas valves will have an input port and a manifold port. If a manifold port is not available on the gas valve, there is one usually located on the manifold.

For natural gas your pressure reading should be 3.5 WC. For LP (Liquefied Petroleum or Liquefied Propane are used interchangeably) your pressure reading should be 10 WC.

If the gas valve does not show the correct reading, remove the aluminum cap located on the gas valve. Using a small screwdriver, adjust the WC pressure by turning the screw in or out as needed until the correct WC pressure is achieved. If you are adjusting an older standing pilot type with a regulator before the gas valve, remove the cap and turn the screwdriver clockwise until the desired WC pressure is achieved. If it bottoms out before you reach the required WC pressure, leave it in the bottom position.

Turn off the gas. Remove the barbed fitting from the gas valve and replace the port plug. Install the barbed fitting on the gas inlet of the valve. This is located before the gas valve. Turn the gas back on and light the furnace. If the gas pressure drops below the required WC pressure, then the outdoor regulator is frozen and/or not working properly. Contact the gas supply company to resolve the faulty regulator.

If the WC pressure does not fall below the required pressure, then the furnace gas regulator or gas valve will need to be replaced. If you have a furnace with a standing pilot, you should replace the gas valve with a combination gas valve and eliminate the regulator. Replace the gas valve on all other models with the specified gas valve recommended by the manufacturer. ***DO NOT REPAIR ANY GAS VALVES…IT IS ILLEGAL TO DO SO.*** Set the gas valve to the required WC pressure. ***Turn off the gas before replacing the plug.*** If you are working on a furnace with a standing pilot, you should perform a carbon monoxide test. If the standing pilot fails the carbon monoxide test, it should be replaced immediately.

CHAPTER 13

Testing Inducer Draw and Negative Pressure Switch

The **Negative Pressure Switch (NPC)** is also known as the **Vacuum Switch**. It is connected by a hose that draws a vacuum from the Draft Inducer Assembly. Vacuum switches have different ranges of vacuum and are used to either pull the switch either open or closed, depending on the needed usage. The **NPC** will have a number on it. The draft inducer assembly is used to vent spent gases as well as draw air for complete combustion through the burner into the heat exchanger. The pressure switch monitors whether this is being accomplished. The **NPC** is used as a safety. If venting becomes restricted or the furnace is not venting properly, the **NPC** will not close, thus shutting off the furnace.

Located on the switch (unless adjustable) is a rating number of WC or In-Ng, which stands for Inches of Water Column. This unit of measurement is usually on a sticker that is attached to the pressure switch. This is the amount of vacuum that is required to operate the switch. By taking the hose from the switch off the Draft Inducer Assembly, you can place a Manometer on the inlet of the Draft Inducer Assembly and turn on furnace. Read the "W.C." Are they enough to pull the switch? If not, check the Draft Inducer Assembly for broken or missing blades or bad gaskets and seals. You may also have to check for restrictions in venting or intakes as on High Efficiency. It is wise to take an Amperage draw and compare it to its rating plate. Sometimes the bearings will drag, slowing the motor down causing an increase in amperage draw on the Draft Inducer Motor. If it is too slow it might not create the vacuum required to close the NPC. If it does draw enough vacuum to pull the switch closed, then you can test the NPC Switch next. To do this, remove the Manometer and reattach the vacuum hose to the Inducer Assembly. Using you DMM set you meter on AC volts. With the furnace off it may not read any voltage until you start the furnace or make a call for heat. Watch your meter as the inducer motor starts it should show 24 volts and then go to 0 volts. If the switch is closed it should be 0 volts. With the furnace running remove the hose from either the NPC or the Inducer. The 24 volts should come back and disappear again when reattached. If it does not, then replace the switch.

Another way to test the switch portion is to put your DMM in continuity and suck on the hose attached to the NPC. When you suck in you should have continuity until you release the air at which time it should go back to an open position. Do not suck on it too hard as you may damage or rupture the diaphragm inside the bellows portion on the switch.

MAKE ABSOLUTELY SURE THE REPLACEMENT SWITCH HAS THE IDENTICAL "W.C." RATING!
Using a switch with a larger "W. C." rating, the inducer may not be able to draw enough to close it. Too small of an "W.C" rating, then the venting could have issues or restrictions causing CO (Carbon Monoxide) to back up into the home without it shutting down the furnace.

CHAPTER 14

Carbon Monoxide Testing

Codes have changed in Michigan for Make-Up Air and Combustion Air to be installed in all new homes, however, the testing for Carbon Monoxide or also known as (**CO**) did not. The old test is now obsolete. The method of putting a detector near a register was the norm. However, with the addition of fresh air (Make Up Air) brought in from outdoors, the readings are diluted. A High Efficiency furnace that brings in Combustion Air is also a dilution. Carbon Monoxide may be diluted enough to show very low amounts or zero amounts of **CO** to register an actual amount at a register on a meter. If the blower motor fails, the **CO** can rise dramatically!

    The proper way to test is to turn gas pressure down to create what is known as a "Dirty Burn." We want to create a carbon monoxide condition or incomplete combustion. After all we are looking for a breach in the Heat Exchanger that may not be able to be seen with the naked eye. Disconnect the blower and allow furnace to fire till it fails from an open Limit Switch. When metal is heated it expands and can open existing cracks in the heat exchanger.

    This allows the metal in the heat exchanger to expand and open any cracks it may have to their fullest potential. Because CO is heavier than air, you will need to push it up for a sample so the meter to read the amount of CO. This is done by placing the meter probe into the plenum (The box above the furnace or the box below the furnace if it is a counter-flow furnace.) Horizontals must be tested at the plenum as well. Turn power off to the furnace. Slowly roll the blower wheel by hand to get the gases to rise. (4-5 turns will be sufficient). Take your reading and follow all allowances for all State, Federal, or Local jurisdictions.

CHAPTER 15

Breaker Trips and Blown Fuses

It is relatively easy to check for the causes of tripped breakers or blown fuses. Before turning the breaker on or changing a fuse, disconnect the largest loads first. On an A/C system, disconnect the Compressor first. **Caution: If removing Power Wires from components or (Loads) be sure to tape them off with electrical tape to insulate them from shorting out on the equipment!** Turn on Power, did the breaker trip again or the fuse blow? Reconnect the compressor and disconnect the Condenser Fan Motor. Reset the breaker or replace fuse and turn on again. Did it Blow or trip again? If the breaker did not trip or the fuse did not blow, the Condenser Motor is bad. The secret is that Control Voltage or Low Voltage loads (24Volt) rarely are large enough to blow a fuse or a breaker even if they are shorted. However, those loads can blow fuses in circuit boards. Just start at the largest loads first and work back to the smallest. Let's say we have a furnace that keeps popping the fuse or breaker. Disconnect the Blower Motor first and work from largest to smallest load. The next possibility may be the Inducer Draft Motor, then the Transformer. Disconnect each item one at a time until it stops blowing fuses or tripping breakers. The last item that causes the fault will be the bad part. Simple isn't it? Hand held breakers may be purchased that can be used over and over indefinitely to save fuses or trips to the breaker panel at most HVAC supply houses.

CHAPTER 16

Testing Flame Sensors

What is Flame Rectification?

When gas is burned it releases billions of electrons from the molecules in the gas. This is referred to as Flame Ionization and it is used as a conductor for electric current to flow from the burner head to the sensor. The flame has a high resistance and is considered a form of a load. The circuit board or ignition box provides voltage. We need a chassis or burner ground to complete the circuit. Conduction can be used to provide a path to close or complete a circuit, therefore providing that a flame was established. Because the sensor is insulated from the chassis ground with usually a porcelain insulator, and the burner is grounded to the chassis, this acts as an open circuit. Because of the difference in the size of the two, meaning the burner is larger than the sensor, the current flows in one direction. When the flame covers the sensor it creates a pulsating DC or (Direct Current) and is now considered rectified. As long as the pulsating DC current is created and the connection is completed, the circuit board or ignition box can hold the gas valve open.

How to Test Flame Rectification or Flame Sensing:

Using your DMM, set it to DC uA or Micro Amperes. It is a good idea to test and ensure that a good ground for the burners established. This can be done by attaching one test lead into the wire and touching the other lead to the burner used for the sensor. Run the furnace in the heating mode. You should have a minimum of .1 uA DC.

Now recycle the furnace with the sensor wire in series with the sensor. In other words, attach one test lead into the wire end and touch the other test lead to the sensor where the wire attaches to it. The minimum reading should be1-4 uA.

If it is not at least 1 the burners will not stay lit. The chemical added to the gas to give it that "rotten egg" smell may have coated the sensor and may have to be removed.

Cleaning a Flame Sensor:

This can be accomplished many ways, but leaving vertical scratches with a 60 grit piece of emery cloth gives the best signal. (Some technicians use a crisp one dollar bill in place of an emery cloth.) Do not polish but wrap the cloth around the sensor and use downward strokes. This leaves more lands and grooves to accept more electrons producing a stronger signal.

If you clean the sensor and the furnace still lights and stays lit for only a few seconds. Check that the burner has a good ground and the board can produce the .1 uA DC. If not you will need to replace the board. It is not likely, but you can check the sensor to make sure the sensor is insulated from ground by putting your DMM in Continuity and making sure there is no continuity between sensor and ground due to corrosion or carbon tracking.

CHAPTER 17

Practice Troubleshooting Series Circuits

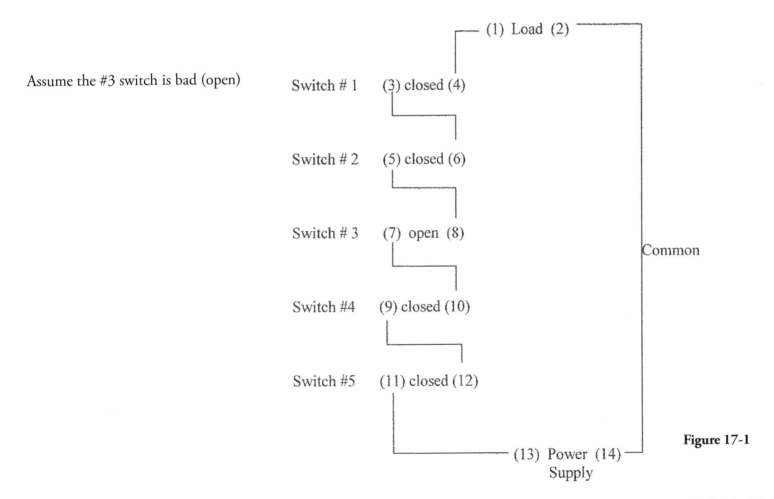

Figure 17-1

Practice test 2

Assume the power supply is bad.

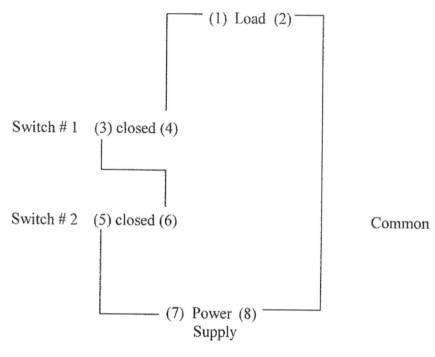

Figure 17-2

Practice test 3

Assume there is a bad load.

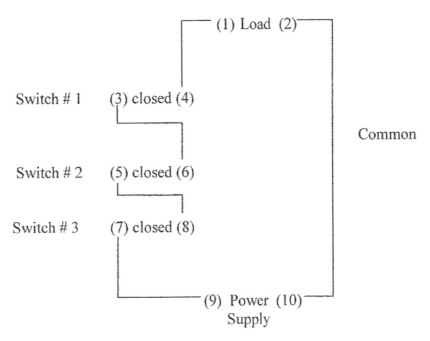

Figure 17-3

Practice Test 4

Assume switch #5 is bad.

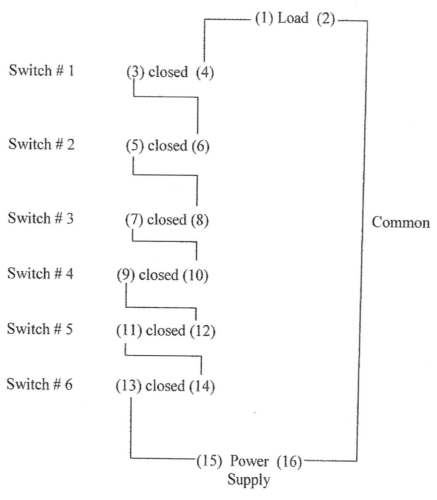

Figure 17-4

Practice Test 5

Assume the common is bad.

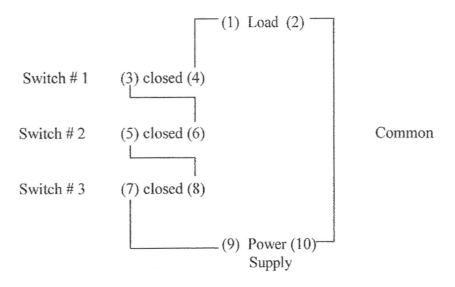

Figure 17-5

Practice Test 6

Practice testing a circuit board. Assume the Blower Motor is bad. The thermostat must be in the heating mode and be above the set point of room temperature. If we have 120 Volts Power at (3) and (5) (Low Speed Blower) and the motor does not run at the proper sequence time, then motor is bad. If we establish that a call for heat has been verified and there is no 120 Volt output at (3) and (5) but we have 120 Volts Power at (4) and (6), then circuit board is bad. Note: To check power on blower motors it will be necessary to trace the wires from the motor back to where they are connected to the board.

This exercise is only an example.

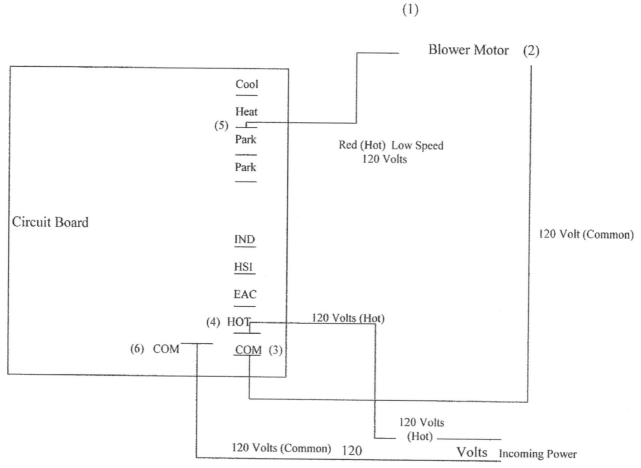

Figure 17-6

CHAPTER 17

ANSWER KEY

TEST 1	TEST 2	TEST 3	TEST 4	TEST 5	TEST 6
1 and 2	1 and 2	1 and 2 power	1 and 2	10 and 1 power	1 and 2
14 and 1	8 and 1		16 and 1	10 and 2 power	4 and 5
14 and 2	8 and 2		16 and 2		See note below
13 and 1 power	7 and 1		15 and 1 power		
13 and 2 power	7 and 2		15 and 2 power		
14 and 1	7 and 8 power		16 and 1		
14 and 4			16 and 4		
14 and 3			16 and 3		
14 and 6			16 and 6		
14 and 5			16 and 5		
14 and 8			16 and 8		
14 and 7 power			16 and 7		
			16 and 10		
			16 and 9		
			16 and 12		
			16 and 11 power		

NOTE: On test 6:
If no power with leads on COM and Hot, then the power supply is bad.
If power is found at 4 and 5, then the motor is bad.

CHAPTER 18

Thermostat Testing

Always check a furnace that does not respond to the thermostat.

Check the power supply on the furnace to ensure the transformer has 24 volts. If it does, we do not need to check breakers or fuses or even the door switch and power switch.

If the house has a heating only furnace with no air conditioning, jump white and red terminals at the thermostat. If it does not fire go on to next step. If when white and red at the furnace is jumped and the furnace runs, then wires from stat, or stat itself is bad. If it does not run, the problem is in the furnace.

If it did run at furnace but not run when jumped at sub-base then wires are bad. Replace wire(s). If furnace operates by jumping white and red at sub-base, then thermostat is bad.

If you have a furnace that has A/C and Heating you can test from red the (R) terminal on the circuit board to the other terminals. Example: Red (R) to white (W) heat should come on. Red (R) to Yellow (Y) Outdoor Condenser should come on. Red to Green (G) the blower in high Speed should come on. If not circuit board is bad. Do not jump Red to Common or the "C" terminal. This will short the transformer and possibly damage circuit board.

If everything works at the furnace then remove the thermostat from the sub-base and jump out the same wires one at a time. Each mode should work. If it does not, the thermostat wire is bad. If it does run then thermostat is bad. If only one mode does not operate at the sub-base but does at the furnace, you may use any unused thermostat wires as a substitute. (Sometimes a brown or blue wire will be unused.) Commonly six wires are used. Otherwise you will have to replace the thermostat wiring from sub-base to the furnace.

CHAPTER 19

Double Series

Many Series Circuits require another Series Circuit to operate. Meaning a smaller voltage series circuit is used to control a higher voltage series circuit. This is true for Relays and Contactors. The method is still the same. We will start with the largest load that does not operate (Sequence of operation) and work back. Then treat the smaller or (control Voltage) the same way. By separating the circuits into two individual circuits it will make it easier to troubleshoot. Look at such a circuit and then break it down to its simplest forms. In this example the blower will not operate because the Coil is bad on the Relay. This coil uses 24 volts to close the N.O. switch to the closed position and allow the 120 volts to flow to the motor. Break it down into two separate circuits as shown in Figure 19-2 and Figure 19-3.

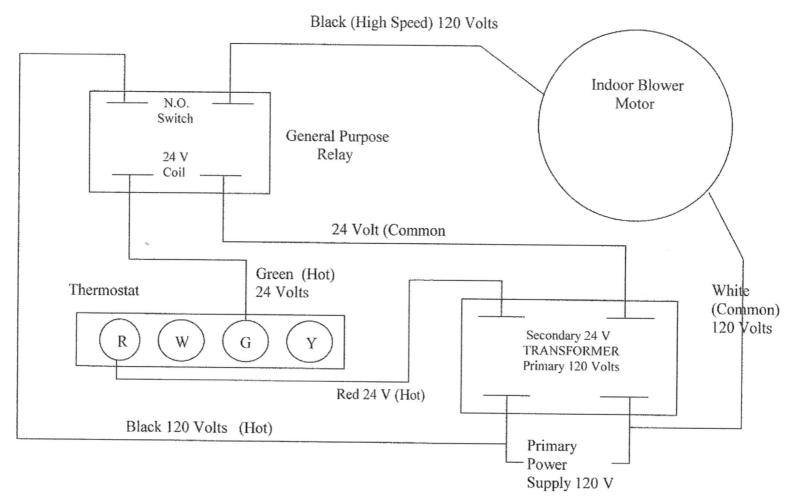

Figure 19-1

First we separate the two circuits as shown in figure 19-2. Notice the 120 Volt portion and the 24 Volt circuits. X marks the test locations. If no Voltage is found, then test 1 and 2

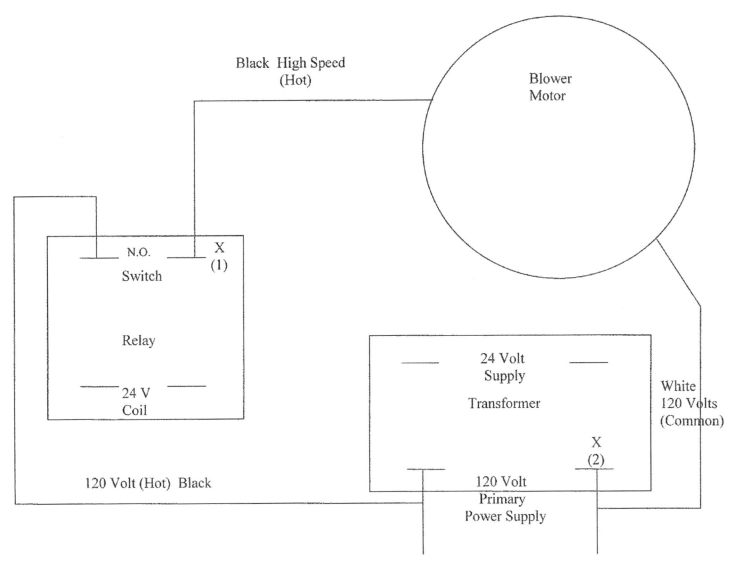

Figure 19-2

Figure 19-3 shows the 120 Volt portion of the double circuit.
When testing 1 and 2, no power is found.
When testing 2 and 3, 120 Volts is found, indicating the switch is open.
Low voltage testing must now be done.

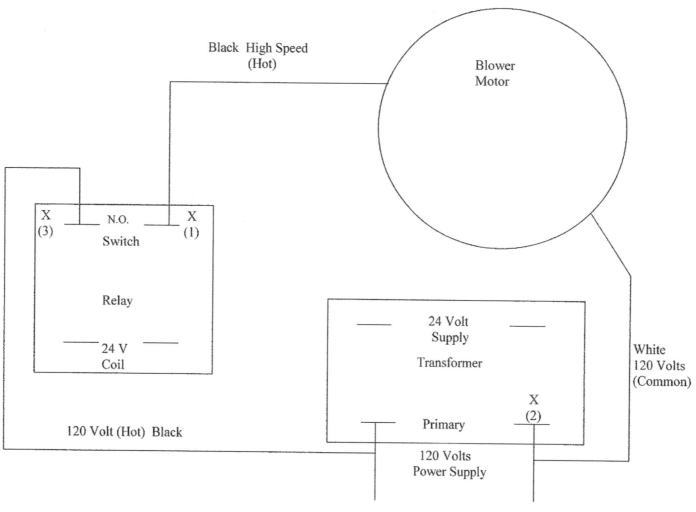

Figure 19-3

Now look at the low voltage portion or the other series circuits which is the 24 Volt control circuit as shown in figure 19-4.

Testing at 1 and 2 indicates we have power but the N.O. switch does not close as in Fig. 19-4, then the coil is bad. Turn the power off and remove the wires from (1) and (2) and test the coil for continuity. If the DMM does not change from "OL" to some form of value then the coil is ("Open"). If power was not present between (1) and (2) the next test would have been from (R) to (G) at the sub-base and if at that time we found 24 volts, then our thermostat did not close. We can also verify power to the thermostat by setting our DMM on AC Volts and check for 24 volts at the transformer. If we do not have Voltage there then test the incoming 120 volts at the transformer. If we have 120 volts on the (Primary Side) of the transformer but the Secondary did not have 24volts, then the transformer is bad. If 120 volts is not found to the (Primary Side) of the transformer then the breakers or fuses will need to be checked.

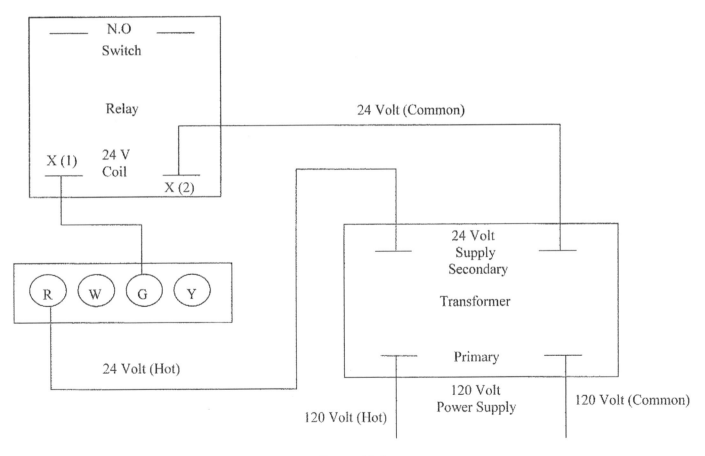

Figure 19-4

HVAC ELECTRICAL FOR IDIOTS | 73

MORE CIRCUIT BOARD TROUBLESHOOTING

In this example the hot surface igniter (HSI) has failed. Figure 19-5 is not drawn to scale. It is drawn this way so the circuit relationships can be seen easily.

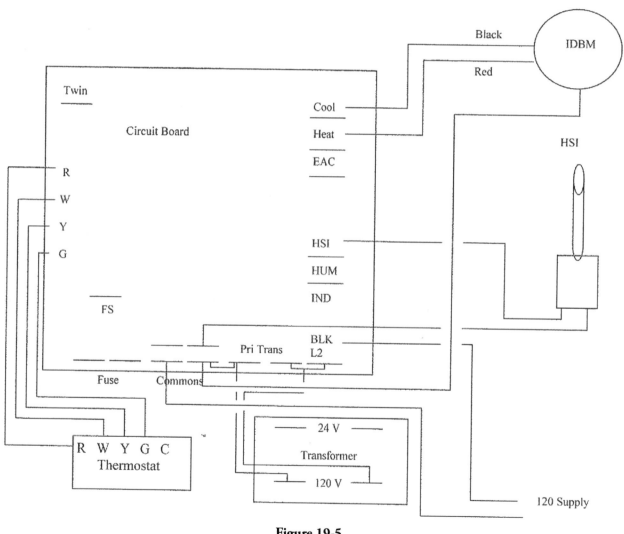

Figure 19-5

MORE CIRCUIT BOARD TROUBLESHOOTING

Now the circuit is broken into its simplest form. Test between 5 and 6. If no power is found, then the part is bad. When power is not found at 5 and 6 and we have verified a call for heat and our sequenced worked up to that point, this indicates that our draft inducer came on and pulled our NPC closed. By process of elimination, it can be assumed the board is bad.

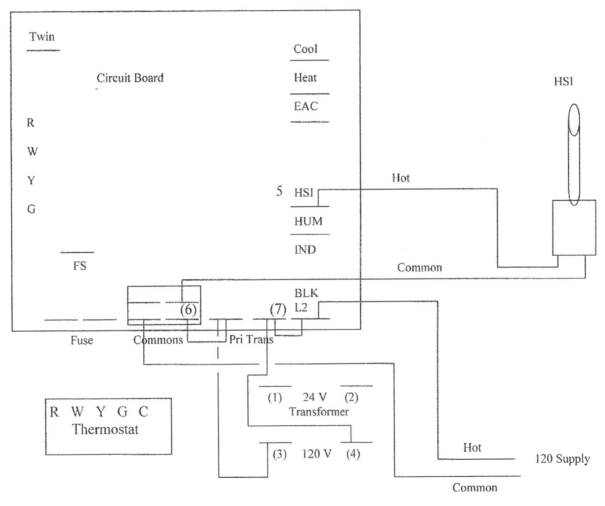

Figure 19-6

CHAPTER 20

Understanding Loops

Figure 19-6 has been reconfigured into the circle format. Look at the troubleshooting process, then break it down to its simplest form. Although the circuits are squared for ease of understanding, treat them as circle circuits. The arrows show the direction of current flow. Fig. 20-2 shows that the Relay Coil is on the Circuit Board.

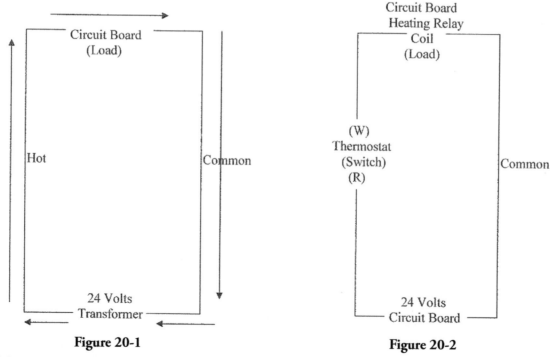

Figure 20-1

Figure 20-2

Figures 20-3 and 20-4 are a breakdown of the circuit board. **IDBM** is the abbreviation for Indoor Blower Motor.

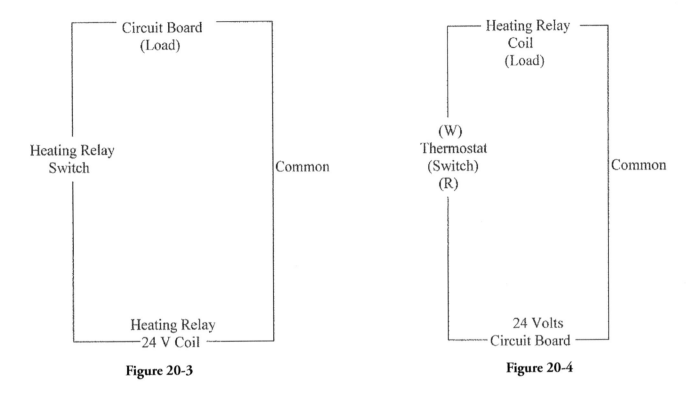

Figure 20-3   **Figure 20-4**

As shown in figure 20-5 and 20-6 each circuit is shown as a loop of its own. It is not necessary to hunt and peck to find the problem. Just identify the loop that is broken and follow the procedures in this book. Remember: Start with the part or Load where the sequence stopped and work backwards from Load to Power Supply. When 24 Volts are applied to the Heating Relay Coil, it then closes the HIS switch allowing 120 volts to go to the Load.

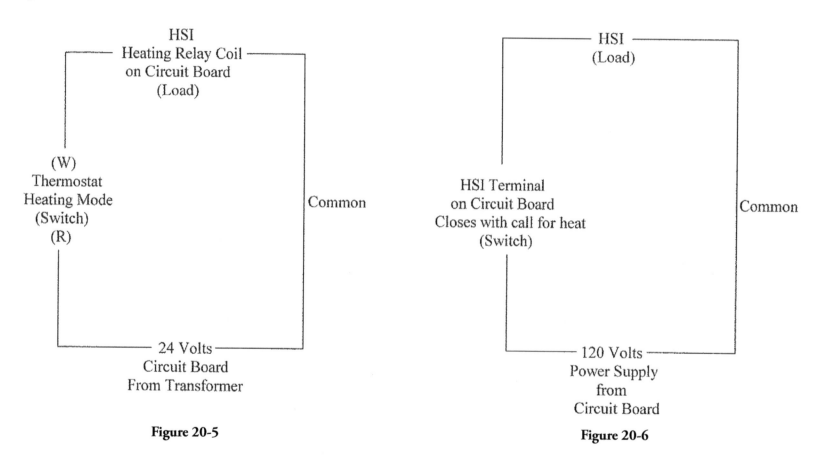

Figure 20-5

Figure 20-6

MORE PRACTICE PROBLEMS

This example is for Residential Air Conditioning.

In this case the Compressor will not run but the Condenser Fan is running. In this situation there is an open Run Winding. There are Two Loads, one from Common to Run and one from Common to Start.

Obviously testing the windings as shown in chapter 10, page 39 would be a faster method to test the compressor. However, the point of this exercise is to demonstrate that testing in the circle of **Power**, **Switch**, **Load**, and **Common** will still work for 230 Volt Single Phase Loads. The Capacitor acts as a start capacitor as it is out of phase by 90 degrees and acts as a bladder that releases its charge when at its saturation point. It then releases its energy as a boost to overcome inertia. As the motor speeds up the current will drop as the resistance is overcome. After that occurs, the capacitor acts as a run capacitor again.

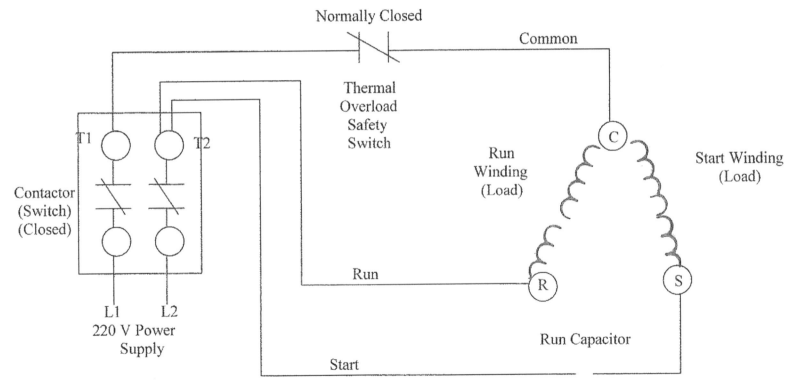

Figure 20-7

HVAC ELECTRICAL FOR IDIOTS | 79

In Figure 20-8, analyze one winding at a time, starting with the run winding. Assume there is an open Thermal Overload Switch.

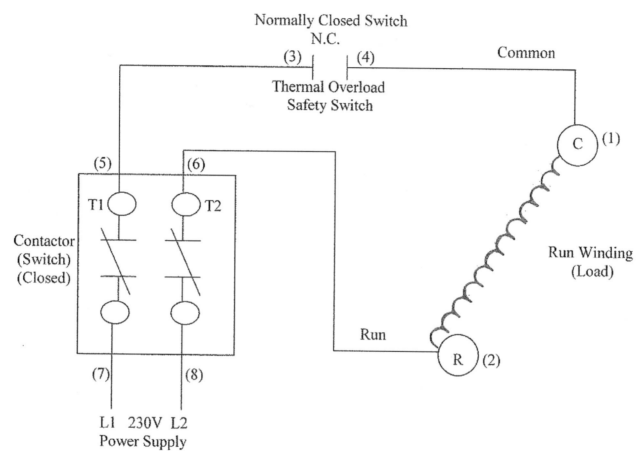

Figure 20-8

Now look at the start winding. In figure 20-8, there is an open start winding.

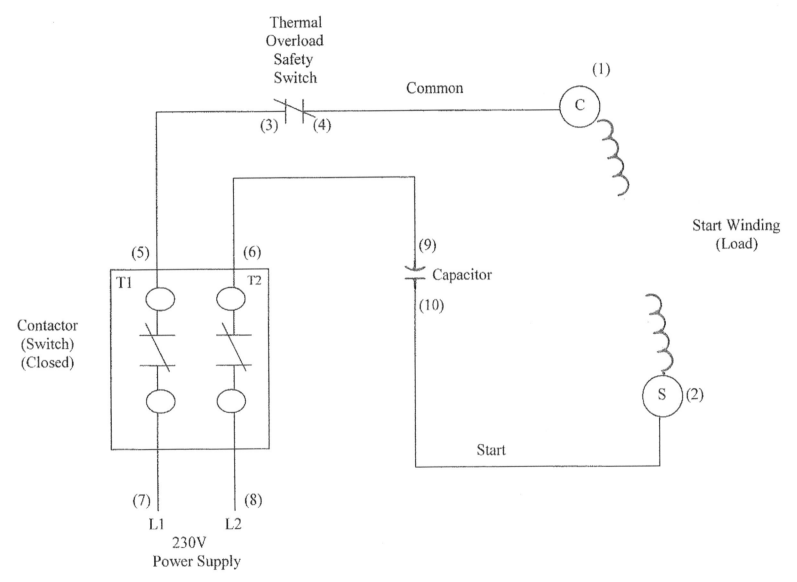

Figure 20-8

SUMMARY

It is easy to see that even with larger voltages troubleshooting procedures stay the same. We just break them down into two separate circuits. With 220,230,240 or any larger single phase voltages, the opposite side acts as the common for the other because of the two cycles being 180 degrees opposite of each other.

The main objective of this book is to show there are two sides to power and you must have both to see the voltage on your meter. Look at the following examples. These examples are just circles with openings or "breaks" in different places.

The first side you see voltage on, you will move to the opposite side to find where the voltage is picked up again. This is due to you changing to the side your missing and supplying it straight to your meter.

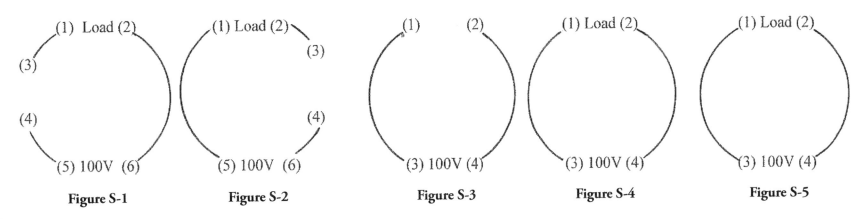

Figure S-1　　　Figure S-2　　　Figure S-3　　　Figure S-4　　　Figure S-5

When we have a complete circuit with a load or some form of resistance, the voltage that is measured at the load will show only from each opposite side. In other words, we will not measure any voltage from the same side of the load, (hot from power supply to hot of load) or (Common of power supply to common of load). Inversely, we will read voltage from hot side of the power supply to the common of the load, and as well we would read voltage from the common side of power to the hot side of the load.

ONE LAST LOOK: Think figure S-6 as a circle but cut in half with load and supply.

The coil representing the load is a wire wrapped around a laminated metal core to form an electromagnet. Electromagnets have many uses and in this case is used operate a relay coil just for illustration purposes. The coil in figure S-6 represents an inductive coil used as power from a transformer. Consider the loop as completely closed, but each side of the transformer is opposite of each other. Voltage can only be measured using both sides of the transformer and any extensions thereof. If we have continuity from each side, the voltage will be able to be measured.

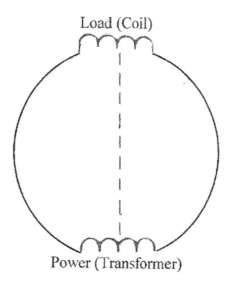

Figure S-6

ABOUT THE AUTHOR

Brien Hollis has been in the HVAC trade for over forty years. He holds several mechanical and boiler licenses. He holds U.S. patent for HVAC equipment and continues to teach HVAC students.

Printed in the USA
CPSIA information can be obtained
at www.ICGtesting.com
LVHW061207171123

764157LV00030B/24